Are We Alone?
The Myth & the Science
Peter Bassett F.R.A.S.

An inter-active book via our website.
www.outerspacebooks.com

Supporting videos and links for this book are offered via the website. Just search the book, followed by the Chapter name.

Astronomy Roadshow Publishing
Further paperback copies, signed copies e-books with direct internet links can be purchased from…

www.outerspacebooks.com

Our Mobile Planetarium service

www.astronomyroadshow.com

Front Cover; the author's image of Sagittarius and UFO model.

Contents

Section One; The Myth?

Section Two   The Science!

Section Three; What Can We Do?

The climax to the movie 'Close encounters of the third kind.' Can this happen or has it?
Image credit; Columbia Pictures

Chapter 1   Introduction

There have been thousands of publications on the subject of UFO's, Aliens, Crop Circles, Conspiracies and the discovery of new planets and moons via various methods. Astronomers generally dismiss UFO's in one sentence, two if they are generous. UFO supporters often dismiss the claims of many scientists just as fast. Conspiracy supporters often portray themselves in a class of their own above everybody else. With the easy publication options on the internet, these subjects have been contaminated and blurred with many wild claims supported with erroneous 'evidence' and claim it as proof.

How do we sort out the mess and get to some kind of truth? With the incredible advances in technology over recent years, it should in theory be very easy. The reality is far from straightforward. With many years' experience in dealing with the fields mentioned above, I have put together this book with the latest discoveries and scientific thinking in mind.

In order to get even close to an answer if we are alone in the Universe or not, we should put aside any personal desire to fall on one particular side of the argument or another. We should ignore all religious viewpoints, disproved discoveries, false reports and invented conspiracies with no substance. To get to the truth, we need to deal with the real world around us in the 21st century. This book handles every aspect one section at a time. It is hoped that it will gradually train your mind into becoming more objective and perhaps encourage some readers to get involved in searching for ETs in a realistic manner rather than wishful or fanciful thinking. Perhaps they do not exist at all, but new technologies and methods are developed for other areas of research as a result of trying.

Some 50 billion species have arisen in one form or another on Earth throughout its entire biological history of over 3 billion years. Only one has ever produced technology along with the thought of the existence of other life bearing worlds. So on Earth intelligent life is a very rare event in its long history.

If we discover that perhaps we really could be alone in this Universe, it would put an enormous responsibility upon our shoulders. We should spread ourselves out across the Cosmos or fail in a quiet puff of meaningless effort in a bland corner of an average galaxy. We should cease arguing over a few acres of land (but the Falkland Islands remain British), stop trying to prove one form of politics is more fruitful than another, or one religion is better than another; and to cap it all cease demonstrating such points through various forms of violence and warfare.

Wasteful habits of the Human Race should halt with immediate effect and go for a global effort to become a long-term thinking species. Through education, technology, and sheer excitement of spreading our influence across the depths of space, we should change our ways and push our talents to higher worthwhile levels. Protect the Human Race first by eliminating poverty, hunger etc., reducing threats from disease, nuclear annihilation and save the environment from ourselves. Then we should have the resources to aim for the big leap to the moon, mars and beyond.

There are no restrictions within the laws of Physics that can stop us heading for the stars one day. Nobody knows how just yet but the rules in nature allow us to do just that. Sir Isaac Newton knew that a Moon landing was possible as far back as the 1680's but had no clue as to how technically it could be done. Today it is a part of our history. Just decades after Apollo 11, we have reached a similar stage regarding interstellar travel, so have other species achieved the same and are even here right now? This book's aim is simply to help us reach the answer to perhaps the biggest question ever asked…

Are we alone?

Chapter 2   Forms of Investigation

In order to solve any mystery regardless of its nature, we have to adopt a structured method. Services such as the police, MI5, the FBI, and CIA etc. all develop methods of approach to their particular field. In science, we need to deal with established facts, mathematics, experiments and realistic possibilities.

The question of whether we are alone or not is always on two levels. The first involves any living creatures at all multiplying happily on their own world without a single thought about the cosmos as a snail does on Earth. The second level is an intelligent species that can build rockets, an Internet system, and spicy pizzas. Both forms will be considered throughout the book but will be more concerned about smart Extra-terrestrials making pizzas.

Every aspect of the possibility of Alien life forming elsewhere in the Universe will be investigated systematically. More than just a sentence or two will be given to UFO's, Crop Circles and conspiracy theories. All angles will be looked at but to the point. Some conclusions can easily be drawn so we can move on toward a more realistic answer to perhaps the greatest quest the Human Race has ever tried to solve.

Alien scene in Baker, California; home of www.ufohotel.com
Photo by the author 2016.

Chapter 3   Ancient Astronauts

Back in the 1960 - 70's a series of books came out by Erich von Däniken regarding the possibility of our civilisation being visited by Aliens thousands of years ago. One of the most successful and influential of all the unqualified archaeologists is this Swiss former hotelier (born 1935). He caused controversy with his popularisation of what has become known as the 'ancient astronaut hypothesis', although he was not the first to propose it.

His first book was 'Chariots of the Gods?' It was originally published in 1967 became a worldwide bestseller, thanks in no small part to its outlook: a written attack on professional historians and archaeologists. This was followed by 'Gold of the Gods.' He was not the first writer to adopt this style and he has certainly not been the last. He is an 'expert basher' and joined ranks with others without a related education. This has become normal practice over the past sixty years or so. The difficulties of Erich's career involved low points of being arrested for fraud, to becoming a global media personality. It ultimately made him very wealthy through selling over sixty million copies of his many regurgitated books. Erich has consistently claimed that the remains of ancient cultures can only be explained by extra-terrestrials within human history and medalling with our future destiny.

The book series included many images and drawings of ancient depictions from Egypt and South America. Many references included phrases such as 'This looks like an astronaut...' 'That looks like an ejection seat...' and so on. Would you regard that as solid evidence of us being visited from far off worlds? Extraordinary claims should be supported by extraordinary evidence.

He once said the massive Easter Island carved stone figures could only have been moved with help from aliens and their powerful flying saucers; that is until a group of students moved such a block of stone by rocking it to and fro with ropes and moved it forward. Erich now admits aliens may not have been required after all. This is where guessing at science gets you; red faced.

Above: Which scene is more likely?

According to ancient astronaut theorists, these figures at Chaco Canyon in New Mexico look like horned aliens carrying a technical device that obviously power their star-ships - so they must be just that then.

Anybody can look at clouds rolling around above our heads and mention the same phrases. This form of 'theory' is a waste of our valuable time. With all this supposed involvement from other worlds over a period lasting many centuries, where are the alien artefacts? Electronic circuits, and unique metal alloys that could only have come from an advanced civilisation from space should be around today for us to discover. All we see are carvings on wood and stone; materials and skills that were widely available at the time. As far as the art is concerned, they are just misinterpreted as aliens carrying gadgets but instead represent something very different.

Ask any real expert in this period of human history and more realistic answers to the petroglyphs are given. The spirals for instance are a representation of the solar cycle; an ever-shortening day and those spirals in the opposite direction represent the oncoming summer instead. An entire book could be dedicated to this massive alteration to true life Archaeology and dismiss such alien myths very easily.

The pictographs above is not of an alien's hand and an impression of its home planet and sun, instead it is of the 1054 AD supernova in Taurus the Bull. Photos by the author at Chaco Canyon in New Mexico 2021.

It has even been suggested many times that Stonehenge in Wiltshire, England was built by aliens too because nobody could work out how it was constructed. Engineers and archaeologists have now agreed how the feat was performed. A group of college students have reproduced a complete plinth placed upon two vertical posts of the same weight and size by hand.

Once again, no aliens were required

The Romans moved two hundred tonne stone blocks often. They built large wheels out of redwood trees, cut out a square hole in the centre, raised the block by a series of wedges and placed the wheels on each end. Just twenty horses or so can now pull the stone block as an axle. Such stones were moved around at Baalbek in Lebanon; no aliens.

Other supporters of ancient astronaut theory include regular TV contributors such as Giorgio A. Tsoukalos of the Legendary Times Magazine and David Wilcock author of *'The Synchronicity Key.'* Neither of these characters has any qualifications in archaeology and yet lay down a completely altered human history based entirely on what some hieroglyphics / ancient drawings may look like. This is not evidence, nor can be considered as scientific. Interesting idea but it is just fanciful nonsense.

Chapter 4   Crop Circles

Many people believe that crop circles have been reported for centuries, a claim repeated in many books and websites devoted to the mystery. Their primary piece of evidence is a woodcut from 1678 that appears to show a field of oat stalks laid out in a circle. Some take this to be a first-hand eyewitness account of a crop circle, but a little historical investigation shows otherwise.

The woodcut was used to illustrate what in folklore is called a "mowing devil" legend, in which an English farmer told a worker with whom he was feuding that he "would rather pay the Devil himself" to cut his oat field than pay the fee demanded. The source of the harvesting is not mysterious — it is supposed to be an image of Satan who can be seen in the woodcut holding a scythe.

According to the original text, the devil *"cut them in round circles, and plac't every straw with that exactness that it would have taken up above an Age for any Man to perform what he did that one night."* This image and story cannot be related to crop circles because it states quite plainly that the crop was cut rather than laid down, as occurs in crop circles.

Some claim that the first crop circles (though they were not called that at the time) appeared near the small town of Tully, Australia. In 1966, a farmer said he saw a flying saucer rise up from a swampy area and fly away; when he went to investigate, he saw a roughly circular area of debris and apparently flattened reeds and grass, which he assumed had been made by the alien spacecraft. Police investigators said was likely caused by a dust devil or waterspout). Referred in the press as "flying saucer nests," this story is more a weather report than a crop circle event.

As in the 1678 mowing devil legend, the case for it being linked to crop circles is especially weak when we consider that the said impression was not made in a crop of any kind but instead in ordinary grass. A round impression in a lawn or grassy area is not necessarily mysterious. Circles have appeared in grass throughout the world that are sometimes attributed to fairies but are actually caused by a crop disease instead. They can also be produced by 'Dust Devil' events; a swirling mass of air rather similar to a tornado.

The first 'real' crop circles did not appear until the 1970s, when simple circles began appearing in the English countryside. The number and complexity of the circles increased dramatically, reaching a peak in the 1980s - 1990s when increasingly complex patterns were produced. These included mathematical and geometrical representations such as fractals.

Crop circles, as the name suggests, usually involve circles — rarely triangles, rectangles, or squares, though some designs contain straight or curved lines. Circle based designs are the easiest for hoaxers to create.

Formations also usually appear overnight, often sighted by farmers or passers-by the next morning. Though there seems no logical reason for extra-terrestrials to only create them at night, it is obviously a great advantage for hoaxers to work under the cover of darkness; large / full Moon nights are especially popular. They never seem to appear on moonless nights.

Camera shyness; 'well they have never been recorded being made' (except, of course, for those created by real hoaxers). This is a very suspicious trait; after all, if mysterious forces are at work, there is no reason to think that they would not happen when cameras are recording. Where are the photographs of flying saucers above crop circles?

Most crop circles show little or no signs of human contact. Hoaxers who devote the time and effort to design and create the (often complex) crop circles are unlikely to carelessly leave obvious signs of their activities such as a Pizza box. They essentially get very good at it. It is their fun pastime.

They also usually appear in fields that have reasonably good public access; close to roads. They never appear in remote, inaccessible areas. The patterns are usually noticed within a few hours of their creation by passing motorists. Any aliens producing them from flying saucers would not need road access. They are also designed to be seen from the air, the excuse of public viewing of Alien evidence from the ground 'so they should be near roads' doesn't make sense either.

There are many theories about what creates crop circles, from aliens to mysterious vortices to wind patterns, but they all lack one important element: evidence. Perhaps one day a mysterious, unknown source will be discovered for crop circles, but until then perhaps they are best thought of as a form of art that annoy farmers.

Responses from this chapter supporting the Alien related theory is expected with vigour. Again, without concrete proof it just cannot be supported as a scientific explanation. I have had comments from my lectures such as *'Well the designs are becoming more complex as a sign that the main Alien mothership is getting closer to Earth.'*

My response is simply *'Well perhaps the press got bored with reporting simple circles, the circle makers had to improve their imagination and techniques to get back on the news.'*

Which answer do you think is more likely?

I cannot see aliens; can you?

To answer the big question proposed by this book, we just cannot constantly fall on the Aliens-being-real side of the debate. It will just produce false hope and conclusions.

The circle makers are very clever at producing such art in rapid time. Just because their full techniques are largely kept secret, it does not mean aliens are involved. We humans have brains that can be creative; aliens are simply not required for such artworks.

To discover exactly how such designs are produced, look up www.circlemakers.org

They have a book out on how to produce such complex circles. The Field Guide: The Art, History & Philosophy of Crop Circle Making by Robert Irving & John Lundberg. Available on Amazon or directly from our website.

Did an alien from a distant star system propose to Laura?

Alternatively; a romantic person who influenced / paid a group to make this.

Aliens are simply not required for such artwork.

Any civilisation that wanted to break the news that we are not alone in the Universe will do so using other methods than crop circles; a system that humans are quite capable of producing with a little thought, planning and team work. Aliens are not required!

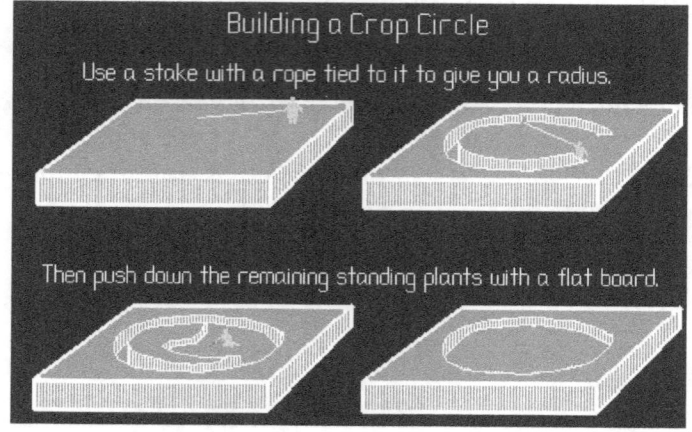

Building a Crop Circle

Use a stake with a rope tied to it to give you a radius.

Then push down the remaining standing plants with a flat board.

16

Chapter 5   The UFO Question

The simplest way of proving if there are Aliens somewhere in the Universe is to look for them in our own skies. Reports have been made for centuries of unidentified flying objects. The vast majority of the human population are not trained scientists. When something unusual is seen, observations are not always made in the same way with scepticism or accurate reporting. Often we hear vague terms of description, speed and height. On occasions an incident was photographed, the details of exposure, camera steadiness etc. also blurs the result in many ways. Such events are rarely observed with a high quality camera on a tripod and shot by an expert photographer. As such UFO events are rare anyway, it is understandable.

Pilots and Astronauts have reported seeing UFOs. Such people have an automatic form of credibility for obvious reasons. Their training includes observing without exaggerating, not to use descriptions that are misleading and so on. In addition, they know that their reputations are on the line if they feel they are miss reporting an anomaly that may already have a simple explanation. They react to such sightings as calmly as possible and report only what was witnessed, not what they wished to witness.

Astronomers are often asked if they have witnessed UFOs. They are a natural group of people to ask such questions. However, most spend their time lecturing in Universities for their wages. Therefore, when they are observing at an observatory on a distant mountaintop, they are looking at a computer screen at a spectrum of a star or co-ordinates of a distant galaxy. Very rarely do they ever observe the actual night sky. Amateur astronomers on the other hand are a different group and are much more likely to witness unusual phenomena. Meteor observers for instance may lie back on a Sun chair for hours at a time counting and recording every shooting star seen. But still they rarely report sightings of UFOs; why is this?

Amateur astronomers understand many forms of natural and artificial night sky phenomena. They recognise regular satellites, satellite re-entry events, meteors, fireballs, aircraft, birds etc. People not familiar with such experiences often report a UFO when it is something perfectly natural or at least explainable.

To make matters worse, some witnesses will often 'add extra information' to their experience especially if anyone claims the sighting to be easily solved. Nobody likes to be embarrassed, so a human aspect has to be considered too.

I have been observing the night skies for over fifty years. Sometimes I have stayed out for up to six hours at a time observing the sky as a whole. Only on two occasions have I witnessed an event that I cannot explain. One was in 1977 when I saw five white lights travelling in a straight line from North to South overtaking each other in a very specific order. Object two overtook object one at the front and became object one. At the same time object five at the back overtook object four and became object three but on the opposite side to the other overtaking event at the front. The process began again throughout the line up in complete silence and was only seen for around twenty seconds. I saw several overtaking events in a seamless loop. It was totally bazaar and have never seen or heard of anything like it since.

The second experience was in 1987 when I was observing the Quadrantid meteor shower in January. I saw three strong red point lights with no meteor-like tail travelling again from North to South. These were all level with each other but slowly spreading apart evenly as if a meteor had broken up and I had just missed the key splitting event; a very rare phenomenon to see in itself.

But then as they reached overhead, the far right object turned sharply and headed due east, the second continued on for 30° or so and headed eastward in a wider arc and the third followed after travelling another 30° in a wider arc still. But the last one faded a little as it passed over some high-altitude cloud that I had been annoyed with earlier; it was reducing the brightness of the third object. This gave an indication of their height. They were not low down in the atmosphere, but high up which meant they must have been travelling at many times the speed of sound to cover such a distance across the sky in just six seconds or so. If any reader has a logical explanation of either event described, please do get in touch.

I do realise that these were just lights in the sky and could not see portholes with green bug eyed monsters peering out; but such reports are bazaar. It only takes one single event with absolute proof of the vehicle/s being occupied by extra-terrestrials to fulfil the answer to the biggest question ever asked. As we have not such proof, I will have to expand this book further and take more of your time; I do apologise in advance.

Fake Reports & Images

The UFO question is constantly marred by false reports and images. False reporting alone without witnesses or any direct evidence is just wasting everyone's time as far as an investigation is concerned. Some do this to gain attention as perhaps they have little in their lives to get excited about. False images however also waste time and damage the credibility of any serious study of UFOs. To demonstrat0e how simple it is I produced my own fake image and we will examine some established proofs of fakery too. Some of which still claim to be genuine by stubborn believers, but trust me, do not waste your valuable time on this Earth with such cases.

This image was produced in 1988. The flying saucer was made of balsa wood and took six months to construct. The lights were optic fibres and the model was superimposed against a darkened stormy sky; perfect for double exposures without letting the background sky 'bleed' into the model part of the image.

Above; Ahah! That is how he did it! The model was supported on a black pole in front of a black cloth background. The model was illuminated internally to give an eerie glow and illuminate the optic fibres. The now vintage computer controlled a stepper motor that rotated the model evenly for a smooth animation; the program was recorded on tape . – All home made.

I produced an animated film of the model rotating against a different background. Such special effects are simple once a few photographic principals are understood.

Proven fake images

Billy Meier, a well-known Swiss writer on Ufology claimed to have taken a series of images of a flying saucer 'orbiting' around a tree. He says these were taken just a few seconds apart. Please observe the cloud formations; would you agree with his statement?

The clouds are in completely different positions in each image. It was obviously a manipulated model and took some time to re-adjust its position between each shot. He should have waited for a blue sky.

The tree was fake too from a model railway store. It was in focus while the horizon was out-of-focus proving the distances were completely different.

An investigator discovered several flying saucer models in a barn of Billy's and his excuse was that he had them made for demonstrations after each sighting of a new flying saucer model. Many people still follow his stories blindly with faith.

To this day Billy insists that these are genuine and the aliens come from the Pleiades star cluster some 400 light years away. Such a young star cluster would be turbulent, swarming with violent actions. Planets are forming around the stars, not yet evolving life for at least another billion years. He should have researched this before making such wild claims.

Enhance the following image digitally and a wire holding up the model is revealed. Flaws can be exposed with many of his 1000 or so images. Given time and full access to the original negatives, each could probably be debunked.

Billy has written hundreds of reports and has a massive global following, but all concerned are wasting their time in this case. I am not claiming that all such images are hoaxed, but such characters make it much harder to get to the truth.

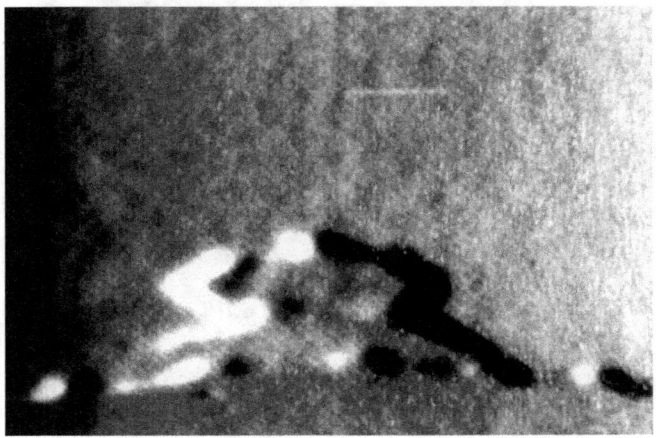

This vehicle contained aliens from the Pleiades Star system some 400 light years away. Billy Meier felt so privileged to be chosen out amongst billions of humans on this planet as an ambassador to the Cosmos...

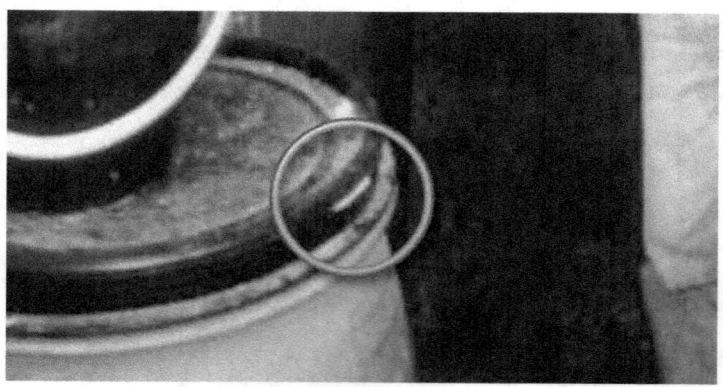

...but then again perhaps the flying saucer was the lid of his water barrel. Even the number of groves and distance between them around the rim match perfectly.

Billy Meier looks happy and content in his work selling books worldwide. I'm sure he is a nice chap if you met him in a bar; but avoid the subject of Aliens.

A complete book can be written on how such images were faked. With digital enhancement software widely available, it is relatively easy to prove whether a flying saucer image is a hoax or not; hence very few are ever published in the papers.

Computer programs can enhance the point that such images should not be taken as evidence at all. Most of the older classical UFO images have since been debunked now as well as the newer. We have yet to see a convincing image that passes all digital tests and detailed enough to be used as absolute proof of the existence of Aliens. Extraordinary claims always require extraordinary evidence.

The TV series *'Alien Files'* has some amazing special effects and interview some very convincing characters, but be aware of their extreme claims. Nick Pope (Britain's UFO expert) he can jump from the story of a Cosmonaut seeing an orange mist outside the window of a space station to concluding it was due to 'angels' or many aliens from different worlds observing them - in one sentence. It was almost certainly due to a venting of gases from any one of numerous valves on the exterior of the spacecraft. At sunrise or sunset (which happens every 45 minutes in orbit) gases outside will glow orange. The Cosmonauts themselves never even suggested an alien presence but Nick Pope did; he invented a mystery out of nothing. He is more likely to fall on the side of a believer and only seems sceptical every so often to *appear* to be objectionable. This book's sole aim is to demonstrate the new and genuine evidence of whether we are alone in the Universe or not.

George Adamski was a classic character that produced many fake UFO images and stories. He claimed that aliens were from Venus and he met the occupants many times and was even taken on a joy ride into space. George claimed to have taken many photographs of a moving flying saucer through his telescope. Have you ever tried to photograph an aircraft through a telescope? Even worse still, he used a Newtonian reflector as shown below. The target will be upside-down and back-to-front in the eyepiece, and so would the motion of the object. Please do try it. It is possible to follow a moving object with such a telescope but require hours of practice. How many hours do we have to follow a flying saucer?

George posing for the photographers in the 1950's. Any amateur or professional astronomer hearing his claim of managing to take many images of a moving target with this telescope will immediately become suspicious.

Left; a confirmed flying saucer from Venus by George Adamski and eventually was offered a lift in it. Ohh do you see the paraffin lamp? Does that look familiar?

The Soviet Venera 4 spacecraft landed on Venus in 1967 and discovered a surface temperature of over 400° C; the almost pure carbon dioxide air pressure was 100 times that of the earth, and rains sulphuric acid. Most of George's followers vanished as fast as his flying saucers. During his lifetime, George sold thousands of books, charged for lectures & interviews and died wealthy. He was not the first and Billy Meier certainly will not be the last to take advantage of die-hard believers.

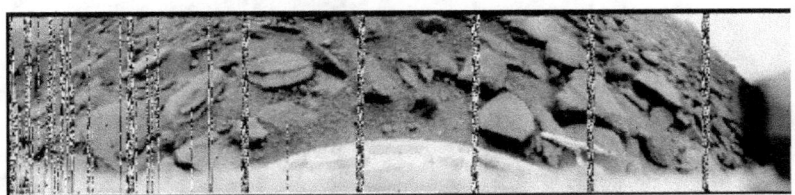

The real surface of Venus imaged in 1975 by the Soviet Venera 9; covered in rocks and lava plains. No aliens here... sorry George.

27

Chapter 6   Roswell

No other UFO event in history has there been more written about than the Roswell case of 1947. There have been books, newspaper articles, US government statements, documentaries and movies based around the incident;

The Roswell Incident
Alien Autopsy
Independence Day
Indiana Jones & the Crystal Skulls
Paul (Arguably the best of the bunch)

Visit Roswell and you will witness museums, alien footprints along the sidewalks, alien-faces on streetlights, a flying saucer shaped McDonalds with aliens, and astronauts hanging from the roof (other restaurants are available). There are also tours on offer to the crash site but be aware, I was reliably informed that it's not the actual site but one with similar surroundings; an interesting experience though regardless.

The Story so far...

On the 4th July 1947, immediately following a massive electrical storm where an almighty explosion type event was heard for miles around, 75 miles from the town of Roswell, New Mexico, a rancher named Mac Brazel found something unusual in his sheep pasture. A mess of metallic sticks held together with tape; chunks of plastic and foil reflectors; and scraps of a heavy, glossy, paper-like material. Unable to identify the strange objects, Brazel called Roswell's sheriff. The sheriff, in turn, called officials at the nearby Roswell Army Air Force base. Soldiers fanned out across his field, gathering the debris and whisking it away in armoured vehicles.

On 8 July, the headlines of the local rag "RAAF Captures Flying Saucer on Ranch in Roswell Region" was the top story in the Roswell Daily Record. However, was it true? On 9 July, an Air Force official clarified the paper's report: The alleged "flying saucer," he said, was only a crashed weather balloon. However, to anyone who had seen the original debris, it was clear that whatever this object was no weather balloon. Some people believed, and still believe, that the crashed vehicle had not come from Earth at all. They argued that the debris in Brazel's field must have come from an alien spacecraft.

To this day, the official U.S. military story claims the debris actually belonged to Project Mogul; a 700-foot-long string of balloons, radar reflectors (for tracking) and sonic equipment. Scientists had launched it from Alamogordo that June and had crashed. Because the project was highly classified, no one at the Roswell Army Air Field even knew that it existed, and they had no idea what to make of the objects Brazel had found. (In fact, some officials on the base were worried that the wreckage had come from a Russian spy plane or satellite –information that they were understandably reluctant to share with the public.) The 'weather balloon' story, flimsy though it was, was the simplest and most plausible explanation they could come up with on short notice. Meanwhile, to protect the scientists' secret project, no one at Alamogordo cleared up the confusion. The Project Mogul 'answer' does not explain much of the activity surrounding the debris recovery. Why the roadblocks and the sending of many large vehicles for a few pieces of tin foil and damaged electronic circuit boards?

The alleged debris; Mac Brazel, as well as other direct witnesses always claimed that this was not the material he had found but was a substitute; the mystery remains. The exchange of debris was clearly shown in the movie 'The Roswell Incident.' USAF image.

Perhaps there is something to the Roswell mystery after all. Again, without solid evidence, it cannot be used to answer our big question. One can believe in the Roswell Incident but that's all it can ever amount to at this point. We need a real alien body, metal alloy, circuit board, or engine component to conclusively solve our quest and not just stories.

Above; the appropriately themed McDonalds and the Roswell UFO Museum research library. Pictures by the author; 2011.

Photos by and of the author taken in Roswell, New Mexico.

Chapter 7   Rendlesham Forest Incident

The UK has had its own major UFO sightings. As the years have passed, the facts concerning one particular case has become hazy as statements change and new witnesses come forward. However, what we do know is that, in the early hours of 26th December 1980, US military personnel (sections of the US Air Force were temporarily stationed at RAF bases in Woodbridge and Bentwaters) spotted strange lights in and above the forest.

One of these men, John Burroughs, accompanied by his supervisor and one other individual, went to investigate reported blue, red, orange and white lights.

In his witness statement published in 1981, Burroughs explains: *"As we went down the east-gate road and the road that leads into the forest, the lights were moving back and they appeared to stop in a bunch of trees... In addition, the woods lit up and you could hear the farm animals making a lot of noises, and there was a lot of movement in the woods. All three of us hit the ground and whatever it was started moving back towards the open field... We got up to a fence that separated the trees from the open field. You could see the lights down by a farmer's house. We climbed over the fence and started walking toward the red and blue lights and they just disappeared."*

Jim Penniston, who accompanied Burroughs into Rendlesham Forest on 26th December, claims to have encountered a craft, covered in hieroglyphic characters. *"I estimated it to be about three metres tall and about three metres wide at the base,"* Penniston later explained. *"No landing gear was apparent, but it seemed like she was on fixed legs. I moved a little closer. I had already taken all 36 pictures on my roll of film. I walked around the craft, and finally, I walked right up to the craft. I noticed the fabric of the shell was more like a smooth, opaque, black glass."*

Burroughs does not recall this. However, indentations on the forest floor, as well as damage to the trees in the area where the lights had been spotted, were found the following morning and support the statement. Radiation levels recorded at the site of the indentations were also unusually high.

In the book 'Encounter in Rendlesham Forest', Penniston writes, *"I left the forest a different man... I was in awe of the technology and yes, a knowing that it was not an aircraft which could have been manufactured in 1980 or even now."*

Both Penniston and Burroughs have since suffered from Post-Traumatic Stress Disorder.

Two nights later, a separate group of military personnel experienced a similar series of events. When the lights were spotted, Lieutenant Colonel Charles Halt was prepared. Halt intended to disprove the wild rumours swirling around RAF bases Woodbridge and Bentwaters. Arming himself with a tape recorder, he set out to investigate. The subsequent audio tape is now considered one of the most valuable pieces of evidence in the Rendlesham Forest incident and is playable on the website... **www.outerspacebooks.com**

Image by the author; 2014.

After an official investigation took place it was decided the flashing light was from a lighthouse on the coast. To this day it remains as the official answer. I may sound a little dim, but wouldn't the same reported lights be visible every single night then? The witnesses were military officials, not prone to exaggeration, nor actively sought publicity, no financial gain and also prepared to risk ridicule; yet the report was still voluntarily made.

This marks the alleged landing spot in Rendlesham Forest. A dedicated book on this incident is 'Sky Crash' by Brenda Butler, Dot Street and Jenny Randles.

This mystery is a fascinating account of a possible alien craft encounter and is still open for consideration in my humble opinion. It is likely that the machine was an early experimental drone being tested in the forest. As it would have been a secret military invention, even the witnesses couldn't be told the truth.

If these lights were indeed from the local lighthouse shining through the winter / leafless trees as the final conclusion claimed, then the same would be visible for months and still would to this day. It is not!

Chapter 8 Snowflake, Arizona

On 5 November 1975, six young woodcutters, along with their employer Mike Rogers, were working in the Apache-Sitgreaves National Forest, carrying out tree-thinning contract for the Forest Service. It is located in east central Arizona fifteen miles from Heber.

The story began at around 6:10pm when the men were heading back home in a crew-cab truck. Travelling along a rough trail, one of them sighted a gold-coloured glow through the woods. Thinking it may have been a crashed plane, they rounded a turn, they saw the source of the glow: a solid object hovering approximately 15 feet above a clearing, ninety feet or so away.

Travis Walton, twenty-two, was sitting on the right- hand passenger side of the front row. When he saw the object, he called to Mike - the driver and boss, to halt. Hardly waiting for the truck to come to a complete stop, Walton jumped out and to get a closer look. As his associates called for him to be careful and come back, he stood and stared at the object, which was at a 60° elevation from his position. It had the shape of two "pie pans" or shallow bowls placed rim to rim. All heard a gentle "beeping" sound.

Walton stepped back, intending to run from it when his friends were startled to see a bluish-green beam shoot out from the bottom of it striking Walton in the chest. It lifted him from the ground with his arms out stretched and dropped him back to the ground seconds later.

Thinking he and the others were in danger, Mike Rogers restarted the truck and left the site. A quarter of a mile away, he stopped and the remaining men looked back. They saw a light rise from the ground and streak into the northeast. Rogers turned the truck around and drove back.

For fifteen minutes, the men searched for Walton, covering the near area and calling him, but no answer. Rogers then decided to drive to Heber and report Walton's disappearance to the sheriff. On the way, they debated what they should tell, doubting that the truth would be believed, but at the same time unable to come up with an acceptable explanation. They chose to tell what they had experienced.

Published in 1978, this book is his own account of the experience he had three years earlier.

On 10 November, the six men were given polygraph tests, which established that they had not harmed Walton themselves. It had been implied that they had killed Travis over an argument and hidden his remains, despite the fact that Rogers was his best friend of many years standing. It also confirmed that they had, actually witnessed a UFO encounter.

On the same night at approximately midnight, a call came in to the Grant Neff residence (Mrs Neff was Travis' sister in Snowflake, Arizona). It was Travis himself, sounding confused and disoriented, saying he was at a phone booth in Heber and in terrible pain. Neff went to Mrs. Kellett's (Travis' mother) home, picked up Travis' brother Duane, who had come up from Phoenix, and drove at breakneck speed to Heber. They found Travis slumped in the phone booth. He had a five-day growth of beard and appeared thin but was otherwise apparently unharmed.

Within hours, Duane drove Travis to his home in Phoenix, intent on keeping him away from the horde of reporters, which had plagued the Walton family during Travis' disappearance, and to obtain medical treatment.

For a short time, the representative of a local UFO group, who sent him to a hypnotist, frustrated Duane Walton. The Aerial Phenomena Research Organisation (APRO) eventually contacted him. They called in a team of medical experts.

Ultimately, Walton was given Lie Detector tests and Psychological Evaluator tests, all of which established that he had told the truth, as he knew it. It was claimed that these tests were conducted and interpreted by experts.

Unfortunately, Walton only recalls an hour or two of his five-day absence. He claims to have awaked on a table in a room, which he first assumed was a hospital. The ceiling seemed low, there was an oval-shaped metallic-coloured apparatus on his chest (his denim jacket and shirt were pulled up), and he was in considerable pain. The air in the room seemed oppressive; warm and humid. It took a few minutes to get his wits about him. When he became fully aware of his surroundings, he realised he was in no ordinary hospital. Around the table on which he reclined were three strange creatures. They were less than five feet tall; pale, with large, domed heads, large eyes, small nose, mouth and ears. They wore tanned orange, seamless thin jumpsuits.

Upon seeing them, Walton struggled to his feet, and when they approached him with their nail less hands outstretched, he grabbed a rod-like object from an adjacent table and prepared to defend himself. After flailing about with the instrument for a moment or two, Walton was surprised to see the trio file out of the door and turn to the right.

After the creatures left, Walton also exited the room, turning left. Following a curved corridor, looking for a way out, he found a circular room with a chair (which was too small for him but he sat in it) with a 'screen' on each arm. He touched a lever and the "stars" on the "ceiling" above seemed to move, so he moved the lever back to its original position and decided against further experimentation.

Shortly, a male, approximately six feet tall, with brown hair and strange golden-brown eyes, appeared at the door, which Travis had entered. He beckoned to Travis, and Travis went to him, babbling question after question, none of which were answered. The "man" said nothing, took Travis by the arm, led him out into the corridor or hall, to the right, and then stopped, whereupon a section of the wall opened. He had not touched anything. They walked into a small room, the door behind them closed, and seconds later a door opened in front of them. They then went down an incline (apparently out of the enclosure Walton had been in) where Walton found him self in a large enclosure resembling a quarter of a cylinder. There were three or four oval-shaped metallic objects parked there (the same apparent metallic substance as everything else he had seen). He was led by the 'man' who was clad in a blue "jumpsuit" with a clear helmet through the enclosure and to another door into a room where there were three other human-appearing individuals-two men and a woman. They resembled the first, although they wore the same clothing, they were without helmets.

They gestured to him to get upon a table. He resisted, but they eventually succeeded and Travis reclined; a device resembling an oxygen mask with a black ball attached was placed over his face and he lost consciousness.

Travis awoke about midnight about a quarter mile west of Heber, Arizona. He was lying on his stomach and he rose up to watch the curved, metallic hull of an aircraft taking off vertically, reflecting the yellow stripe of the dividing line of the highway below.

What did Travis Walton see? What did he experience? Tests seem to indicate that he has related his experience truthfully. His book The Walton Experience (1978) tends to illuminate the reader and enable them to make their own judgment.

39

As with all such cases, there is another side to the story. All seven guys involved always claimed that they never wanted to profit from the experience, just report it. At the time, the logging business that Travis and Mike Rogers owned had financial difficulties. None of the crew took the work very seriously and lacked motivation.

Military helicopters were often on exercise in the area where they were working. Training missions included search and rescue in forests using bright search-lights. Perhaps an idea for a hoax UFO encounter was born from this.

Fortunately for the 'witnesses' the polygraph (lie detector) tests were not conducted by experts, but by people who had limited knowledge on how it worked. Questions such as 'Did you see a bright light coming from the sky?' were asked. This may have been very true as there was a helicopter on a training exercise instead of a craft containing aliens. Not a lie there so far. The wrong kind of questions was being presented. Another later test showed he was lying completely and the interviewer said 'it was the plainest case I've seen of lying seen in 20 years.'

A TV crew from CBS demanded real experts and a different result. Psychiatrists were chosen and put Travis through a long session of analysis. Their methods were certainly unique. The next day the four of them disappeared into a room and soon a waiter headed there with two bottles of cognac.

At the end of it the psychiatrists were rolling drunk but they had their story. It seemed that Travis' father had deserted his family as a child, but had been a spaceship fanatic and all his life, Travis fantasised riding in a spacecraft.

Travis Walton in the foreground showing off his cheque for $2,500 for his story to the Enquirer newspaper. The other witnesses had a share of a further $2,500.

Travis had seen something out there in the woods, some kind of an eerie light which had triggered a powerful fantasy. There does not seem to be any kidnap by aliens.

Reports began to filter in that the witnesses' lie detector tests were not much help - they supported the story that they had all seen the strange light but not necessarily a spaceship. The movie based on the case *'Fire in the Sky'* is shown in a different light and should not be referred to as an accurate portrayal of what happened. The names and dates are the same except the logging crew showed five members rather than seven.

To this day all seven witnesses involved though have stuck to the same story. This is certainly unusual as many people regurgitating a fantasy story tend to change it and exaggerate as time passes. Where does that lead us?

The Sitgreaves National Forest; close to the alleged abduction site, a few miles outside Heber, Arizona. Photo by the author 2017.

There is unfortunately no solid evidence of a true abduction by aliens in this instance, but had come close to being taken seriously by some scientists. All seven witnesses do charge and accept payments for interviews, book signings etc. and only agree to such deals with organisations that support the existence of UFO's rather than sceptics. With the last points becoming known, sadly, this is another story that cannot be used to answer our big question.

A wood carving store at Heber sells 'aliens' as a visible tribute to the Travis Walton story in the region. Travis gives talks about his story every November in Heber, Arizona. Photo by the author; 2017.

Chapter 9   Navajo Reservation, Arizona

In northeastern Arizona a rural town named Chinle, was buzzing with the news that on 20 January 2009, a UFO had landed on a remote ranch. Rangers from the Navajo Nation Department of Resource Enforcement reported that a Navajo elderly couple and a girl saw a UFO land near their house in a rural area northeast of Chinle. According to the ranger, the alien occupants of the object went up to their house and used what could be described as narrow beam of light to look around the house and the outhouse before leaving.

The Navajo descendant walked outside to meet them without any fear. They exchanged glances and the beings calmly walked back to their craft and flew off into the starry night. An area of flattened grass was reportedly found at the site of the landing the next day. The family never sought publicity, financial gain or any other form of reward for their report.

Its cases like this one that remains intriguing, simply because there was no known motivation for reporting such an amazing story. The Navajo calls the beings 'Sky People.' The family kept the same story to this day and remain consistent from all three witnesses.

Millions of reports have been made for decades but very few stands up to scrutiny. The vast majority are easily explained away, some are just clear lies, but a small number of interesting cases do remain.

Chapter 10   Lonnie's Close Encounter; New Mexico

One of the most intriguing cases of a UFO sighting with a little trace evidence is that of the 1964 landing of a craft witnessed by police officer Lonnie Zamora. This event has been one of the key marks in Ufology for decades.

The incident began at 5:45pm on 24 April 1964 in Socorro, New Mexico. Thirty-one-year-old experienced police officer Lonnie Zamora was on patrol when he was passed by a car, which was obviously speeding. He took off in pursuit but suddenly heard a loud roar in the distance, accompanied by a bluish, orange flame rising into the air.

He knew that there was a dynamite shack not too far away and he thought at first that there had been an explosion there. Abandoning the chase for the speeder, he drove his police car in the direction of the shack. He radioed his activities to the sheriff's dispatcher.

As Zamora proceeded towards the rising smoke and flame, the aftermath of the explosion seemed to disappear and reappear because of the rising and dipping roads he travelled. The route he was on was a narrow gravel trail and it wound around a small gully. As he approached the location of the shack, he noticed in the distance a shining object, around 100 to 200 yards away.

His first reaction to this sight was that it was a car which had overturned, and its gas tank had exploded. Upon a closer look, he discovered that it was an oval-shaped object without windows or doors. He stated that the object was about the same size of a medium-sized car.

He was drawn to an unusual red insignia on the side of the object, and then noticed two beings that he thought at first to be children, dressed in white overalls. He recalled that one of the 'children' seemed to jump to attention upon noticing him.

Regaining his composure, Zamora immediately radioed the sheriff's office the details so far witnessed. He decided to get a closer look at the strange scene before him. He then heard a loud roar, and a bluish flame shot out of the underside of the object. Afraid that it was going to explode, he dived to the ground to protect himself, bushes in the vicinity caught fire.

Next, he saw the object lift off the ground, and head southeast, flying in a straight line for about 10 miles then flew upward. The legs that he had seen earlier had disappeared rapidly into the craft. Having intercepted Zamora's earlier radio transmission, State Police Sergeant Sam Chavez arrived at the scene just after the craft disappeared high into the sky.

Lonnie Zamora never sought fame or fortune and risked his career reporting this incident. He had no known motivation to invent such a story.

Following extreme ridicule by his colleagues, he took early retirement from the Police Department.

On 25th April, the first military investigator on the scene was Army Captain Richard T. Holder, Range Commander of White Sands Missile Base, along with an FBI agent, D. Arthur Byrnes, Jr. from the Albuquerque office. Major William Connor from Kirtland Air Force Base and Sgt. David Moody representing the Air Force Project Blue Book on UFOs met Lonnie the next day. The famed Dr J. Allen Hynek arrived on the 28th. Hynek conducted a follow-up investigation in August 1964 to see if his story had changed as many hoaxers do; it had not. The following is an excerpt from Capt. Holder's report:

"Present when we arrived was Officer Zamora, Officer Melvin Katzlaff, [and] Bill Pyland, all of the Socorro Police Department, who assisted in making the measurements. When we had completed examination of the area, Mr. Byrnes, Officer Zamora, and I returned to the State Police Office, Socorro then completed these reports. Upon arrival at the office location in the Socorro County Building, we were informed by Nep Lopez, Sheriff's Office radio operator, that approximately three reports (from the public) had been called in by telephone of a blue flame of light in the area... the dispatcher indicated that the times were roughly similar..."

The following document from the CIA was made available for public inspection in 1981.

"Studies in Intelligence," released in 1966. Hector Quintanilla, Jr., the former head of Project Blue Book, wrote the brief, "Policeman's Report,"

"There is no doubt that Lonnie Zamora saw an object which left quite an impression on him. There is also no question about Zamora's reliability. He is a serious police officer, a pillar of his church, and a man well versed in recognizing airborne vehicles in his area. He is puzzled by what he saw, and frankly, so are we.

"This is the best-documented case on record, and still we have been unable, in spite of thorough investigation, to find the vehicle or other stimulus that scared Zamora."

The case received a great deal of press, and a lot of attention by UFO groups around the globe. The single negative aspect of the Socorro incident is that Zamora, though considered honest and reliable by everyone who knew him, was the sole witness of the event up close. However, do not forget three individuals from the public phoned the emergency services that they too saw a bright light on the ground in the distance at the same time Zamora was at the site radioing the details.

Zamora took such ridicule from members of the police force and local community that he retired only two years after the reported incident.

The case again does not prove the existence of extra-terrestrials, but there is almost no doubt that some type of craft with occupants did land and take off. Dr J. Allen Hynek, who interviewed Zamora on more than one occasion, believes every word by Zamora but no explanation for his sighting.

In Hynek's own words; *"There is much more evidence to indicate that we are dealing with a most real phenomenon of undetermined origin."*

If what Zamora saw was not of extra-terrestrial origin, then where did it come from? Why did it land? Who were those small occupants?

During several visits to the cowboy town of Socorro, I have met three unrelated people who knew Lonnie for many years before and after the incident. They all mention how he was modest, quiet, and never enjoyed any publicity. He just wanted a peaceful life in Socorro amongst his family and friends. He had no motivation to invent such a fantastic story. Events such as this one that demonstrate an open mind should remain on this subject and not dismiss it completely.

Harold Baca in 2016, resident of Socorro for decades knew Lonnie Zamora personally. Harold's son, Shane, was so inspired by the incident that he studied Astrophysics at New Mexico University. He acquired a technical post at the Very Large Array – a Radio Telescope just 80 miles from Socorro.

Drone shot of Zamora's mural near the alleged UFO landing site – Socorro, New Mexico Aug 2022.

Chapter 11   San Augustine Crash

A not so well known potential UFO crash site is around 150 miles to the West of the Roswell site, this alleged crash occurred around the same time in July 1947. The key witness was Grady (Barney) Barnett of Socorro. A debris field was discovered and reported by Grady; he was a soil expert and was surveying the area at the time. Nobody knows the exact date or time of this incident; there were no witnesses, just the debris discovery.

The exact location has remained a secret, but keen UFO hunters have discovered the site and mysterious pieces of material have been recovered. The exact composition of the samples is still unknown. Stanton Friedman has written books on this site such as 'Crash at Corona'.

Above; Grady's modest home in Socorro; right photo, now abandoned (author image 1992).

Today, the Plains of San Augustin is the home of the most sophisticated radio array on the planet – the Very Large Array. Some Ufologists claim that it was built there to communicate directly with ET. The site in reality was chosen for its 'radio quiet' properties, flat terrain for construction, low-cost land and many abandoned local railway sources for some of the construction materials. No conspiracy here, sorry to sound boring.

This book is just to give an overview of the UFO mystery. Most scientists dismiss it with a wave of the hand, but more serious attention should be given to UFO's as it would be the easiest and lowest cost method of discovering our quest if true. Our research facility in Arizona is a small step toward this goal... www.northstaroasis.co.uk

The VLA on the Plains of San Augustin.

This monument in the heart of Socorro is known as 'First Contact' - a very relevant name for this area. It includes a representation of the VLA.

Chapter 12   Police Chase in Ohio

On the morning of 17th April 1966, police officers Deputy Sheriff Dale Spaur and Mounted Deputy Wilbur "Barney" Neff was investigating an abandoned car on a country road called Route 224 of Portage County, Ohio. They gave chase to a UFO after asking advice on the police radio. They first saw the object rise up from near ground level, bathing them in light, near Ravenna, Ohio at about 5:00am. They were ordered by the sergeant via radio to pursue the object. They chased it at high speed for eighty-five miles across the border into Pennsylvania. Along the route, police officers from other jurisdictions saw the object and joined in the chase.

"I always look behind me so no one can come up behind me. And when I looked in this wooded area behind us, I saw this thing. At this time, it was coming up...to about tree top level. I'd say about one hundred feet. It started moving toward us.... As it came over the trees, I looked at Barney and he was still watching the car...and he didn't say nothing and the thing kept getting brighter and the area started to get light...I told him to look over his shoulder, and he did.

He just stood there with his mouth open for a minute, as bright as it was, and he looked down. I started looking down and I looked at my hands and my clothes weren't burning or anything, when it stopped right over on top of us. The only thing, the only sound in the whole area was a hum...like an overloaded transformer.

I was petrified, and, uh, so I moved my right foot, and everything seemed to work all right. And evidently, he made the same decision I did, to get something between me and it, or us and it, or whatever you would say. So we both went for the car, we got in the car and we sat there...."

As they watched, the UFO moved toward the east, and then stopped again. Spaur reported the movement to the dispatcher.

The UFO was now about 250 feet away, and was brilliantly lighting up the area *("It was very bright; it'd make your eyes water,"* Spaur said.) Sergeant Schoenfelt, off duty at the station, told them to follow it and keep it under observation while they tried to get a photo unit to the scene.

Spaur and Neff turned south on Route 183, then back east on Route 224, which placed the object to their right. "At this time," said Spaur. *"It came straight south, just one motion, buddy, just a smooth glide..."* and began moving east with them pacing it, just to their right at an estimated altitude of 300-500 feet, illuminating the ground beneath it. Once more the UFO darted to the north, now left of the car, when they had to speed up to over 100 mph to keep pace with it.

As the sky became brighter with predawn light, Spaur and Neff saw the UFO in silhouette, with a vertical projection at its rear. The object began to take on a metallic appearance as the chase continued. He kept up a running conversation with other police cars that were trying to catch up with them. Once when they made a wrong turn at an intersection, the object stopped, then turned and came back to their position.

Police Officer Wayne Huston of East Palestine, Ohio, situated near the Pennsylvania border, had been monitoring the radio broadcasts and was parked at an intersection he knew that the Portage County officers would be passing soon. Shortly afterward, he saw the UFO pass by with the sheriff's cruiser in hot pursuit. He swung out and joined the chase. At Conway, Pennsylvania, Spaur spotted another parked police car and stopped to enlist his aid, since their Cruiser was almost out of gas. The Pennsylvania officer called his dispatcher.

According to Spaur, as the four officers stood and watched the UFO, which had stopped and was hovering, there was traffic on the radio about jets being scrambled to chase the UFO, and *"...we could see these planes coming in...When they started talking about fighter planes, it was just as if that thing heard every word that was said; it went PSSSSSHHEW, straight up;*

and I mean when it went up, friend, it didn't play no games; it went straight up."

The Air Force claimed the UFO was a satellite, seen part of the time, and confused with the planet Venus. Under pressure from Ohio officials, Major Hector Quintanilla, chief of Project Blue Book, had intense confrontations with the witnesses and refused to change the identification, although it was pointed out to him that they had seen the UFO in addition to Venus and the moon. Major Quintanilla also denied that any jets had been scrambled. Satellites only look like moving stars with no beams of protruding light.

William B. Weitzel conducted an exhaustive investigation on behalf of the National Investigations Committee on Aerial Phenomena (NICAP), obtaining taped interviews, signed statements, sketches, and all pertinent data, which was assembled into a massive report.

This incident inspired the UFO chase scene in the movie *"Close Encounters of the Third Kind"*. J Allen Hynek recommended it to Steven Spielberg himself. This is a case with several witnesses, police officers from different areas who did not know each other before this event; a very troubling case for non-believers.

Chapter 13   Spaceship Moon?

We have covered Ancient Astronauts, Crop Circles and UFOs, what about *'Our Moon is hollow and is a massive spaceship'*? Yes, you heard it right. There are a few strange facts about our natural satellite. Firstly, it is light compared to the Earth; the Volume of Earth is 50 times that of the Moon but is 81 times heavier. The Moon has one side that faces the Earth all the time, ideal for aliens to observe us for years.

The Moon formed because of a glancing blow to Earth by another planet early in the history of the solar system. The iron on Earth had already sunk to the centre, so had the planet's iron that hit us. The iron core of the intruder sank to the earth's core and merged; only lighter materials formed the Moon after impact. The moon's side that always faces us is a phenomenon known as 'Captured Rotation.' The interior mass of the Moon is slightly uneven so eventually the heavier side slowed down its spin until it synchronised with the earth. All these unusual points have been explained.

We then hear of pictures of stone towers, cities, and spaceships on the moon. The birth of the internet gave such people who believe this a 'voice.' This is just a short explanation of where these people are going wrong.

Thousands of buildings.

'*These structures look like rectangles and squares on the Moon so they must be buildings.*' Alternatively, is it that this blurred picture from a low-resolution camera is enlarged so much that the pixels from the CCD picture show and reveal the shape of them instead?

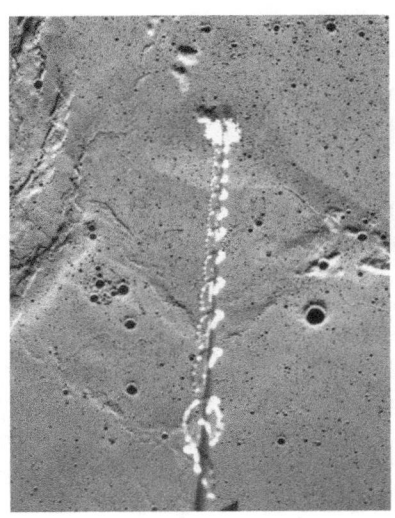

'*This is a tower several kilometres high and is standing on the moon.*'

Alternatively, is it a 2mt long antenna attached to the orbiting spacecraft that is on hundreds of other pictures?

'The shadow below is far longer than the crater shadows so it must be a tower rather like an Egyptian Obelisk. It must be one of those or a mobile phone mast for Aliens.'

Alternatively, is it a normal rock standing on the edge of a downward slope so the shadow will be longer for that natural reason instead? Let us get real please.

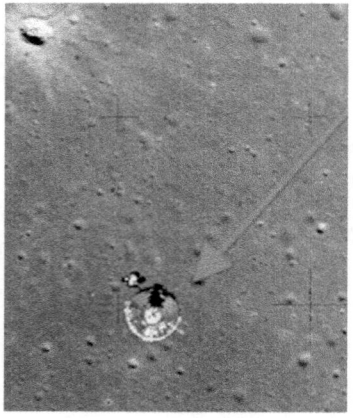

'Here we can clearly see an alien base on the moon's surface.' Actually, we are looking at the top of the Apollo 14 Command Module from the Lunar Module window.

You get the idea by now. I will not waste any more of your irreplaceable time.

Chapter 14 The Shape-Shifters

There is a subsection group of UFO enthusiasts that believe our civilisation has been infiltrated by a reptilian species that can change their body shape, texture of their skin and even the shape of their eyes. According to British conspiracy theorist David Icke (yes him again), tall, blood-drinking, shape-shifting, reptilian humanoids from the Alpha Draconis star system are hiding in underground bases on earth. He claims that such creatures are the force behind a worldwide group controlling various historic human events.

Icke contends that most of the world's leaders are, or at least related to these reptilians; including the Queen of England, President Barac Obama, the Pope, and President George W Bush (perhaps they might be onto something after all). These conspiracy theories now have supporters in many countries and David Icke has given lectures to crowds of up to 6,000 people. American writer Vicki Santillano included it in her list of the 10 most popular conspiracy theories, describing it as the *'wackiest theory'* she had ever encountered in her global research.

This picture is the absolute proof that the queen of England is a reptile from the star Alpha Draconis... Ohh really?

The first appearance of such reptilian aliens in literature was in Howard's story, *'The Shadow Kingdom'* published in August 1929. This story drew on other ideas of the *'Lost worlds of Atlantis'* and another publication *'The Secret Doctrine'* with its reference to 'dragon-men.' These were a mighty civilization on a fantasy continent. They were described as humanoid with human bodies but snakeheads, able to imitate any human at will. They lived hiding in underground passages, using their shape changing and mind control abilities to infiltrate humanity. Episodes of Star Trek, Space 1999 and X-files included similar fanciful creatures. My favourite movie reflecting this subject is *'Sleepwalkers'* by Stephen King in 1992 and then there was a children's TV series called *'The Tomorrow People'* that also included the idea. These are all works of fantasy only.

Many members of the public seem to take in this conspiracy without stepping back and take in a reality check. For those that are absorbed by this crazy hypothesis, there are special hospitals and treatments for help.

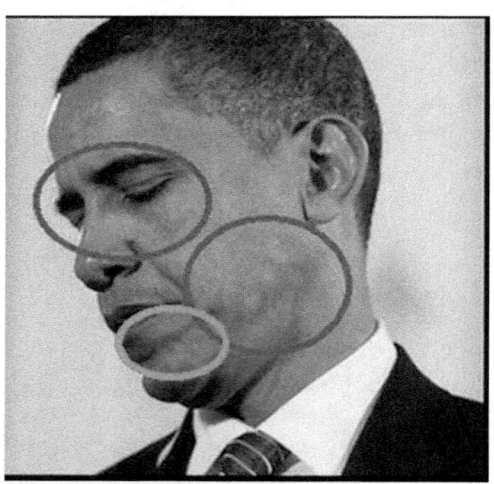

Above; very obviously, Barac Obama is a reptile???

This hypothesis does not deserve any more than two pages.
YouTube clips to laugh at are on www.outerspacebooks.com

Chapter 15   UFO Fraud at Aztec, Mew Mexico

The well-known alleged UFO crash site of Roswell has had a competitor that has now largely faded into history; Aztec. Roswell benefited financially as the hotels, bars and restaurants filled with UFO chasers, reporters and tourists in general. Aztec has always struggled to make its own mark on the New Mexico map.

Author Frank Scully published a story in 1949 that a 30 metre (100ft) wide flying saucer had landed hard with sixteen humanoids on board. The alleged site was in the Hart Canyon, sixteen miles northeast of Aztec; a very remote area that could only really be reached on horseback or jeep at the time.

His sole source of information was from two travelling sales representatives; Silas Newton & Leo Gebauer. They claimed the military closed in and extracted craft and bodies in secret. Later the pair began to sell a miraculous device that could detect Gold, Oil and gas deposits deep underground. They claimed that some of the technology was based on an Alien source. Geologically, the area is indeed rich in Oil & Gas but not too much gold.

Frank Scully believed the wild claim and published it often in his books and in the *Variety Magazine* that specialised in anything a little different.

The story gave birth to the Aztec UFO Symposium, which was run by the Aztec library purely as a fundraiser from 1997. As a severe lack of evidence became more apparent, the Ufologists lost interest and the group had its final convention in 2011.

Today, the story within Aztec is regarded as a light hearted piece of history. It has become an attraction for UFO enthusiasts, even though the truth behind it has been widely accepted.

Chapter 16   UFO Treatment

The cases I have brought to the reader's attention are based in the UK & USA. The only reason for this is that I have been to the alleged site(s), personally met an investigator who worked on the case, met the witnesses themselves or at least friends of theirs. I could not see the point of mentioning case after case that I read from other sources and regurgitate the same material without showing a new angle.

In many countries around the globe, the number of unexplained UFO reports is largely increasing. This is despite the fact that there are better-trained investigators to weed out the explainable cases almost immediately. Over the years, I have encountered several investigators from the Mutual UFO Network (MUFON) and the British UFO Research Association (BUFORA). They now seem to be very down-to-Earth people who are deeply interested in the subject but do not shout 'Aliens' at every report. Many of them have a firm understanding of astronomical, meteorological and aviation related phenomena to explain away most UFO reports.

I have taught a few investigators about unusual events that are caused by satellites and mistaken as UFOs. When they re-enter the earth's atmosphere, they break-up and produce a colourful display of glowing material flowing across the sky, some fading out, others brightening. It just depends upon the individual melting point of the various materials and the shape of each piece producing a different amount of air resistance.

Another new phenomenon is Iridium Flares. There is a series of satellites (Constellation of satellites) that have large gold-plated flat antennae on board. These Iridium satellites reflect sunlight as a mirror does and produce a reflected beam of sunlight on the Earth around 10 miles wide travelling along at 5 miles a second. The result is that an observer on the ground will see a flash of light lasting for around 2 seconds then nothing. Such light flares are so bright; they can even be seen in daylight. I have mentioned them in more detail along with my own recorded clips on www.outerspacebooks.com

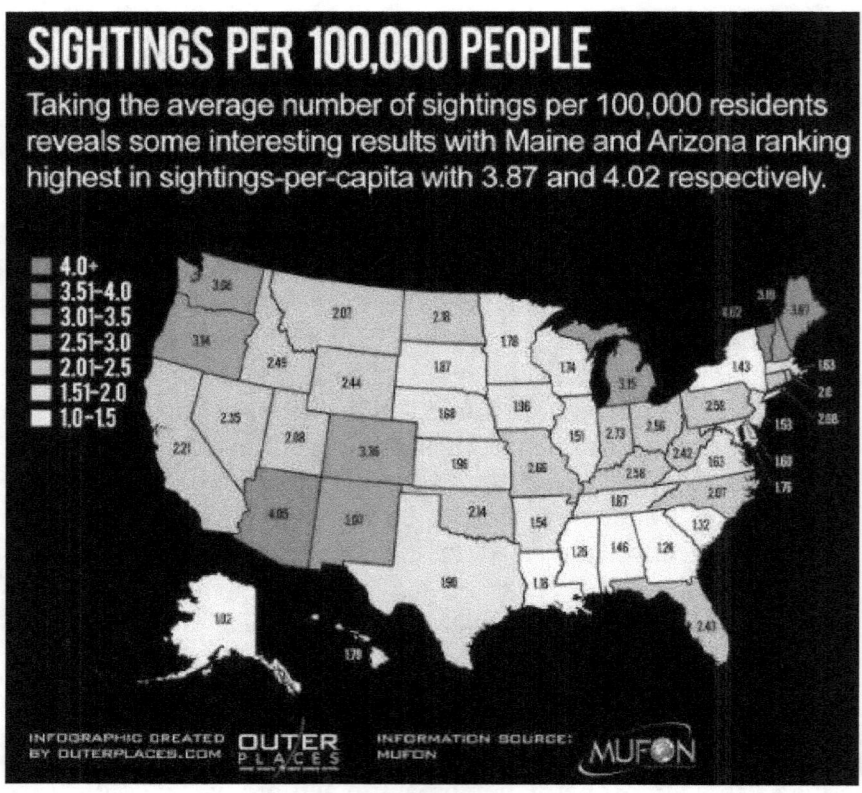

The graph above show the number of UFO reports against population size in the USA. Clearly, the highest reporting coincides with the darkest skies. Arizona is at the peak, hence a reason for setting up our own observing site there; www.northstaroasis.co.uk

Within the first few nights of observing we produced our first case; now being listed on the MUFON database as this book is released. (UFO Investigations chapter).

Chapter 17   The Open UFO Conclusion

During a lifetime of examining UFO cases via books, TV documentaries, chatting to witnesses etc., the vast majority are easily explained away. However, the human desire to reach a conclusion that these things are real is so strong, that many people are ignoring the realistic answer to a sighting or even a close encounter with aliens and believe the fantasy entirely.

Example; to the inexperienced observer, satellite re-entry events are often reported as UFO's. On the 1 August 1974, dozens of witnesses including two prominent UFO 'experts' David & Barry Biedny made a sighting over Venezuela. They estimated their UFO to be around 1.5km away.

A true analysis showed that this was a part of a rocket that was launched from Baikonur, Kazakhstan on 29 July that year. It was a Proton vehicle third stage and was left in a decaying orbit of 51°, 182 x 198km. It was 6.5m long, 4.15m in diameter and a mass of 3,500kg. It was assigned the international designation 1974-060B and a US Strategic Command catalogue number 7393. The actual distance to the sparkling debris was around 300km not 1.5km as estimated by the 'experts.' Mystery solved! It was tracked on radar and Russia admitted to the re-entry event. But David & Barry still insist to this day they witnessed a real UFO and still have a dedicated 'Alien Files' episode on TV.

Dots are Flying Saucers now?

Many clips on You-tube show a dot moving in front of the Moon and the publishers scream 'Aliens' without hesitation and yet to an experienced astronomical observer, it's clearly a satellite. Such attitudes give the whole study of UFOs a bad reputation and little wonder that scientists with a wave of a hand often dismiss the subject. Dr Carl Sagan was one such well-known character.

If UFO investigators were more prepared to accept a scientific explanation of a 'mystery' especially when there are no doubts left to answer, then perhaps the subject could be taken more seriously in general.

However, there are a tiny proportion of cases that defy all investigative methods. The cases that we ought to take a particular interest in are those witnesses that sought no publicity or financial reward. Such reporting may result in a potential damaging mark on their careers and reputations and yet kept the consistent storyline. The Travis Walton case of Snowflake, Arizona was close. Even though there was no photographic evidence, all the witnesses concerned hardly changed the story. The Rendlesham Forest incident is a perfect example of a sighting that could be taken as genuine even though it could have been an early drone aircraft. The Zamora episode is another example.

Many other cases are not documented for various reasons. One took place in the 1950's near Oak Grove, California. A farmer saw a large round object touch down on his land one night for a while and just flew off. Since then, nothing grew in that patch for many years. Not having any interest in science, he carried on his daily work without giving any official report, but the story remains amongst the local population.

Beauty Peak; the alleged scene of an unreported UFO landing; west of Oak Grove, near Warner Springs, California.

Dr Carl Sagan held a strong objection to the UFO subject *'Why do all major sightings that involve a clear view of aliens always seem to occur in the middle of nowhere instead of showing themselves fully above a city?'*

My answer is the same as NASA's policy on the exploration of Mars or any other body that may have life on it. Whenever a space probe is intended for a landing where life could be present, then that craft has to be cleaned from every microbe possible before launch. Upon landing, there should be as little disturbance as possible to reduce the risk of altering the habitat of that world. Any contamination could alter the entire future and studies of that planet.

Any creatures visiting us too could apply the same policy. They may wish to study us as a species but not to alter the entire future of our race. If such a vehicle hovered over Washington DC, as in the movie *'The Day the Earth stood still'* then in an instant, we would know we are not alone in this universe. Political, military and religious groups will be running around wondering what to do next. The entire course of our civilisation would alter beyond recognition in a day. It could never return to what we had before. So actually, Carl Sagan's biggest objection can be answered using the very organisation he worked at for most of his career.

There seem to have been many more convincing UFO cases from the 1940s to the 1980s but fewer since. Perhaps this is because it is now so much easier to expose fraudulent cases by digitising images and enhancing them to see how the image/s was produced. As there are so many quality cameras around, there should be better quality images produced of this phenomenon. Alternatively, perhaps, the extra-terrestrials know that our recording technology is so extensive, and radar systems now so powerful, that that now leave us alone more. Both arguments do make practical sense.

Therefore, I have to leave the UFO phenomena open in my mind. This would by far be the easiest way to answer the question of whether we are alone in this Universe or not. Some people say they have actually seen beings from another world walk around right in front of them and fly off as with the Chinle, Arizona and Socorro, New Mexico events. Such witnesses, if genuine, will already have the answer I am seeking. However, without the concrete proof at present that science desires, we cannot use the subject yet for a definite conclusion. It may keep the believers satisfied, but is not scientific and cannot be accepted; a great shame though.

The science approach...

When Galileo built his telescope and looked at Jupiter in 1610, he discovered four worlds orbiting it rather than the earth. He could prove that only the Moon truly orbited us and we were not at the centre of the universe as claimed by the church at the time. Anyone can have their own telescope and see the same. Newton came along and demonstrated how each planet and Moon moved by mathematics and made predictions from those calculations. Again, anybody could perform the maths, make predictions in the future positions of the planets and they would work perfectly. Einstein did the same by expanding those laws to take in extreme speeds and masses.

To this day, everything works perfectly from very small atoms to high speed planets and super massive stars and black holes. All the branches of the Theory of Relativity have been tested now to an incredibly high level and still stand.

Such pursuits can be tested and verified by anybody with the right mathematical and technological skills. Regarding the subject of Flying Saucers, we have not been able to verify their existence at all. Without a piece of material, a biological sample, clear images that stand up to testing etc. we just cannot as yet use this subject to solve our mystery.

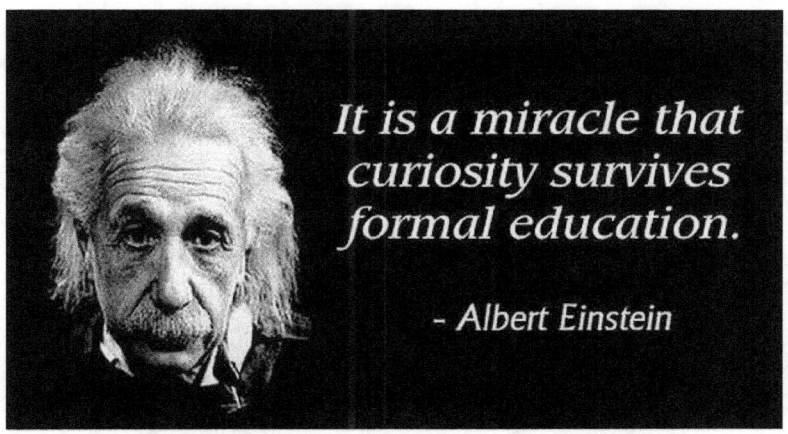

It is a miracle that curiosity survives formal education.

- Albert Einstein

We now have to take this question to a different level and involve hard scientific investigations that can be tested and repeated. This will include subjects of Biology, Chemistry, Astronomy and potential targets to explore within the Solar System and beyond. Future forms of Spaceflight will also be examined along with the potential lifespan of any civilisation such as ours.

Each section will represent a part of arriving closer to our goal; the knowledge of whether we exist alone in this vast Universe or have the potential of qualifying to join massive community millions of light years across. Perhaps this book is a universal constant and that every single species such as ours ask exactly this very same crucial question until the answer is revealed.

The entire Human Race once felt that the Earth must be at the centre of the whole Universe as it looks as though everything is indeed revolving around us every 24hrs. Then Galileo proved the Sun was at the centre instead and Newton showed that Gravity controlled the Planets, not God (although it can be said that God defined Gravity 13.8 billion years ago; that works fine). Then it was realised that the Solar System was not at the centre either, and by the 1920's, nor was our Milky Way galaxy in prime position.

Religious groups of varying denominations tried to fight against this scenario for centuries and many still do. So as when most teenagers become rather demoralised when they discover that the whole world is not made just for them, we have the same outlook through science that the whole Universe wasn't made just for us.

It would be so easy to accept this concept but without the proof, it would be irresponsible to draw conclusions either for or against. Artwork by friend Richard Green – 1988.

Section 2
The Science

The author posing in front of the world's second largest steerable radio telescope at Greenbank, West Virginia; the site of the first scientific search for alien transmissions.

Chapter 18   The Frank Drake Equation

Frank Drake was born in Chicago on 28 May 1930 and was raised in Chicago's South Shore. He had a typical childhood but had a constant interest in science. He spent hours with friends experimenting with motors, radios, and chemistry sets. He had a particular interest in astronomy and began to wonder about the possibility of the existence of other planets and the potential life they could harbour. The idea seemed reasonable to him. However, because of the religious convictions of his parents and teachers he never felt comfortable bringing up the subject.

After high school, Drake enrolled at Cornell to study electronics. It was here that he became obsessed with astronomy and in 1951; he attended a lecture by Otto Struve, one of the world's greatest astrophysicists. Towards the end of a lecture, Struve showed that there was mounting evidence that planetary systems had most likely formed around half of the stars in the galaxy. He went on to state that life could certainly exist on some of those planets. Finally, Drake had found someone who shared his ideas.

After college, he spent the next three years with the Navy to repay his scholarship fees. Thanks to his degree, he ended up as the electronics officer on the USS Albany where he gained invaluable experience operating and fixing the latest high tech radio equipment.

When his Navy service ended, Drake headed to Harvard graduate school to study optical astronomy. Fortunately, the only summer position available was in radio astronomy. Because of his experience in the Navy, he was a natural fit because the radio equipment was constantly in need of repairing and upgrading. It was here that Drake was hooked on radio astronomy and never looked back.

Upon finishing graduate school in 1958, he got a position at the new National Radio Astronomy Observatory (NRAO) in Green

Bank, West Virginia. From here in 1960, the first serious science based search for alien signals took place. Named Project Ozma by Drake, the search was a two-week observation of the stars Tau Ceti and Epsilon Eridani. At one point during the search, a false alarm turned out to be a normal terrestrial signal, but did cause some excitement at the time. He hardly expected to find evidence of advanced civilizations on the first search, but it had to begin somewhere.

Dr Frank Drake

In 1961, Frank Drake and J. Peter Pearman, from the Space Science Board of the National Academy of Sciences, organised the first Search for Extra-terrestrial Intelligence (SETI) conference. The three-day meeting, held at the NRAO, was a small gathering of a dozen or so scientists who had shown an interest. It was in preparation for this conference that Drake came up with the now famous Drake Equation.

$$N = R^* \times f_p \times n_e \times f_l \times f_i \times f_c \times L$$

N = the number of civilizations in The Milky Way Galaxy whose electromagnetic emissions are detectable.

R*= the rate of formation of stars suitable for the development of intelligent life.

fp = the fraction of those stars with planetary systems.

ne = the number of planets, per solar system, with an environment suitable for life.

fl = the fraction of suitable planets on which life actually appears.

fi = the fraction of life bearing planets on which intelligent life emerges.

fc = the fraction of civilizations that develop a technology that releases detectable signs of their existence into space.

L = the length of time such civilizations release detectable signals into space.

The next few chapters will deal with each factor in turn with the very latest evidence to support the figure. When the formula was first used, almost none of the components were defined to enable anyone to reach any conclusion; it was just a wild guesstimate. As our technology improved, particularly since 1990 with the launch of the Hubble Space Telescope, several of the figures are now defined within a maximum and minimum range with a high degree of confidence.

A major adjustment also need to be taken into account; if a civilisation develops a form of travel between the stars for beings or machines, the chances of detecting them increase to more than just the number of planets they evolve from. In addition, a colony does not even have to live on a planet; massive space stations are possible that can house thousands of occupants for generations. Such vehicles will need radar systems and communications with fellow beings many light years away; another opportunity for detection by us nosey humans.

Any advanced civilisation does not necessarily require a planet to live on. Transmissions can be made between massive stations and planets, increasing our chance of detecting them.

To calculate your own 'number of civilisations' in the Milky Way, use the link on <u>www.outerspacebooks.com</u>
From the corresponding chapter name.

Chapter 19   Conditions for Life

So what makes a world such as ours able to host life? Why is Earth so special? There are a few key ingredients that scientists often agree are needed for life to exist. Discussions remain with many scientists as to what limits there actually might be on life. Even our own planet hosts some strange creatures that live in extreme environments.

Here is a simple rundown of what makes life able to thrive on our home planet and likely for alien life to arise on other planets and moons.

Water: First, you would need some kind of liquid, any place where molecules can mix and react. In such a soup, the ingredients for life as we know it, such as DNA and proteins, can swim around and interact with each other to carry out the reactions needed for life to develop.

The most common substance is the one life uses on Earth: water. Water is an excellent solvent, capable of dissolving many materials. It also floats when it is frozen, unlike many liquids, meaning that ice can insulate the underlying water from freezing further. If water instead sunk when frozen, this would allow another layer of water to freeze and sink, and eventually all the water would freeze, making the chemical reactions for life impossible.

Astronomers looking for extra-terrestrial life most often concentrate on planets in the habitable / goldilocks zones of their stars — orbits that are neither too hot nor too cold for liquid water to persist on the surfaces of those worlds. Earth happened to hit the Goldilocks mark, forming within the sun's habitable zone. Mars and Venus lie on the outer edges; if the Earth's orbit had been just a bit further inside or outside of where it is, life may likely never have arisen and the planet would be a cold desert like Mars or a cloudy furnace like Venus. Of course, alien life may not play by the rules we're used to on Earth, but should.

Astro-biologists increasingly suggest looking beyond conventional habitable zones. For instance, while liquid water might not currently persist on the surface of Mars or Venus in large quantities, it might have evolved on their surfaces in the past, and then fled to safer underground places or adapted to the environment when it became harsh. Extremophile organisms have on Earth adapted to living in acid or boiling water. In addition, other liquids might host life; Saturn's Moon Titan has liquid methane and ethane capable of harbouring the processes for life as far as we know.

Energy: Second, life needs energy. Without energy, virtually nothing would happen. The most obvious source of energy is a planet or moon's host star, as is the case on Earth, where sunlight drives photosynthesis in plants. The nutrients created by photosynthesis in turn are what the bulk of life on Earth directly or indirectly relies on for fuel.

Countless organisms on Earth live on other sources of energy such as the chemicals from deep-water vents. There may be no shortage of energy sources for life to thrive on.

Time: Scientists have argued that habitable worlds need stars that can live at least several billion years, long enough for life to evolve. Some stars only live a few million years before dying, we can therefore eliminate such stars for our search to save time. . Life might originate rapidly, but for complex organisms such as those we see today require at least a billion years of steady environments.

For instance, the Earth is about 4.6 billion years old. The oldest known organism first appeared on Earth about 3.5 billion years ago, meaning that life might conceivably evolve in 1.1 billion years or less. However, more complex forms of life did take longer to evolve — the first multi-cell creatures did not appear on Earth until about 600 million years ago. Because our Sun is so long-lived and stable, higher orders of life, including humans, had plenty of time to evolve.

Recycling Planets: Some researchers suggest that it is vital for a world to host life; it must have an outer layer that is broken up into plates that constantly move around and recycle the air and land.

For instance, carbon dioxide helps trap heat from the Sun to keep Earth warm – the greenhouse effect. This gas normally is bound up in rocks over time after being absorbed either chemically or biologically via photosynthesis in trees etc. If this continued, then the planet would inevitably freeze. Plate tectonics helps this rock to be dragged downward in subduction zones, it melts, and this molten rock eventually releases the same carbon dioxide gas back into the atmosphere through volcanic eruptions.

Mars has no tectonic plates but does have volcanoes. This would allow a higher build-up of carbon dioxide, which would keep the planet warmer than normal for its distance from the sun. As it is a much smaller planet than earth, the core cooled quickly and volcanic eruptions ceased. Eventually the atmosphere was lost due to its weak gravity.

There may be other sources to emit carbon dioxide to keep the balance, or perhaps the planet could be a little closer to the parent star to prevent a global freeze in the first instance. This feature of Earth may not be essential for life after all, but helpful in our case.

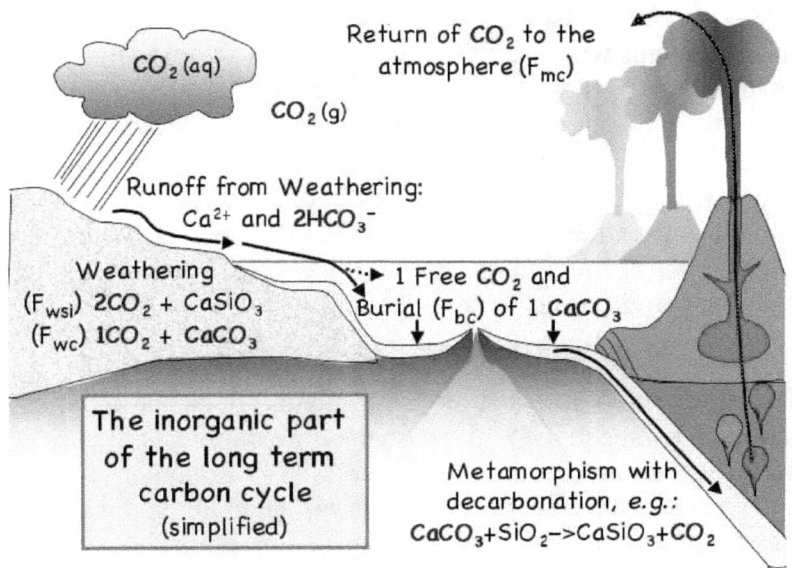

CO_2 (aq)

Return of CO_2 to the atmosphere (F_{mc})

CO_2 (g)

Runoff from Weathering:
Ca^{2+} and $2HCO_3^-$

Weathering
(F_{wsi}) $2CO_2 + CaSiO_3$
(F_{wc}) $1CO_2 + CaCO_3$

1 Free CO_2 and
Burial (F_{bc}) of 1 $CaCO_3$

The inorganic part
of the long term
carbon cycle
(simplified)

Metamorphism with
decarbonation, e.g.:
$CaCO_3 + SiO_2 -> CaSiO_3 + CO_2$

Above; Carbon Dioxide is absorbed by trees and plant life in the oceans. This material becomes buried in the land. Millions of years pass, some of this land is pushed downward in a 'subduction zone.' It heats up due to interactions with magma and can be re-released in a volcanic eruption.

Extra features: Researchers have trotted out for why life succeeded on Earth include how little variation there is in our sun's radiation compared with more volatile stars, or how our planet has a magnetic field that protects from any storms of charged particles from the sun. Violent bursts of radiation could have mass extinctions on Earth in its early, fragile stages.

Earth remains the only known planet to host life due to a unique combination of factors. However, continued monitoring of other worlds might one day change that, by finding other planets that share these attributes or by discovering other ways that life has found to blossom elsewhere in the Milky Way and beyond.

Extremophiles (Lovers of Extreme Environments): These are organisms with the ability to thrive in environments such as hydrothermal vents or pure acid. Since they live in high pressure and temperature, they can tell us under which range of conditions life is possible. The unique enzymes used by these organisms, called 'Extremozymes,' enable these organisms to function in extremely forbidding places. These creatures hold great promise for finding life on other worlds as it widens the search to environments that as humans, we are not acclimatised to.

It's important to note that these organisms are bazaar only from a human perspective. While oxygen, for example, is a necessity for life as we encounter it in our parks, some organisms flourish in environments with no oxygen at all. Here are some of their categories;

Acidophile: An organism that grows best at acidic (low) pH values.

Alkaliphile: An organism that grows best at high pH values.

Anaerobe: An organism that can grow in the absence of oxygen.

Endolith: An organism that lives inside rock or in the pores between mineral grains.

Halophile: An organism requiring high concentrations of salt.

Methanogen: An organism that produces methane from the reaction of hydrogen and carbon dioxide, member of the Archaea.

Oligotroph: An organism with optimal growth in nutrient limited conditions.

Barophile: An organism that lives ideally at high water pressure.

Psychrophile: An organism with optimal growth at temperature 15 °C or lower.

Thermophile: An organism with optimal growth at temperature 40 °C or higher.

Hyperthermophile: An organism with optimal growth at temperature 80 °C or higher.

Toxitolerant: An organism able to withstand high levels of damaging elements (e.g., pools of benzene, nuclear waste).

Xerophile: An organism capable of growth with very little water.

This is a Bacterial Mat at Yellowstone National Park, Wyoming. Microscopic organisms are living quite happily in a solution similar to battery acid at almost 100° C Acidophile / Hyperthermophiles. Photo by the author 2014.

Various forms of Extremophiles live on earth. Caves systems are one such example. Cut off from the outside world, these vast volumes often filled with water, can host many forms of life that are non-existent above ground. Some of the longest and deepest caves can take an experienced caver several days to reach the end. Some biologists specialising in this area has made many discoveries and widens the habitats that living organisms can exist in.

Cathedral Caverns near Huntsville, Alabama. Various forms of Extremophiles have been discovered at the very end of such caves where human presence has not disturbed the living conditions as yet. Some caves are specially preserved for scientific research and are kept sealed or at least remain secret.

The largest cave in the world is in Vietnam and has had fewer human visitors than the Moon for preservation reasons.

The Water Bears / Tardigrades: A tiny creature the size of a full stop is a very remarkable species. The 'Tardigrades' have been discovered to be incredibly tough and very long lived. It is classed as an animal as it has limbs, claws, a brain etc. It is the only animal so far discovered that could survive the harsh environment of space; it can handle radiation, extreme temperatures, and no air pressure for years. Many have been taken up on the International Space Station and exposed to the outside in orbit. They all survived and fully recovered back on Earth after long duration exposure to space.

It is now known that the DNA strands in Tardigrades are coiled up so tightly, that radiation traces miss them constantly.

This unlikely product of evolution has resulted in a creature that is almost indestructible as an individual and as a species. It is currently not known how long these have been around for on earth, but if ever we suffered a massive extinction event on earth, then creatures such as these will be ensuring a continuation of evolutionary lines.

This demonstrates that as such life forms evolve; the planet / Moon should harbour life until a total destruction of the habitat concerned due to the star expanding to a red giant or a similar event.

Past Massive Changes: Since the formation of Earth, massive changes to our climate have indeed taken place for various reasons. One of the most studied of recent finds is of a global ice age that resulted in the entire planet being covered in a layer of ice. It is now thought that around 800 million years ago, so much plant life existed in the Oceans and land, that it reduced the Carbon Dioxide (CO_2) content of the atmosphere to almost zero. At the same time, volcanic eruptions also settled to an all-time low period of activity. Such events normally kept CO_2 levels stable and kept the planet reasonably warm.

The result was a cooling of the atmosphere, land temperature, and commenced a global big freeze. The snow and ice reflected more sunlight back into space and cooled our world further. Eventually the entire globe was encased in ice for millions of years. Life on land and air became virtually extinct. The oceans suffered too with only a few species surviving to continue on the evolutionary tree.

At some stage tens of millions of years later, the volcanic activities began again and increased the CO_2 content. The ash from eruptions covered the ice with dark material to absorb sunlight and begin to melt the ice and accelerate the process. Life then repopulated the land and diversified much further than ever before.

The point being made here is that we may find worlds completely frozen or even locked into the opposite direction, but if life had taken hold in the distant past, then it does its very best to adapt and survive, even if it exists deep under the ice or encased in rock. As conditions change, it thrives again on a larger scale than before. We must not look at planets or moons as they are today and make a decision of whether they are life-bearing objects, but examine the past too. They may seem hostile today, but living creatures in the form of Extremophiles for instance may still occupy that world.

Chapter 20   Suitable Stars

We naturally think of our star the most suitable for providing life on Earth. It is very stable i.e. no massive regular flares or large changes in energy output. It has also to be very long lived; around 10 billion years. This would give plenty of time for life to develop and change into beings that are more complex and eventually produce some form of intelligence. So how common are stars such as our own?

Stars that form from clouds of Hydrogen gas alone are called Population I type stars. Any planets that evolve around them would be much like Jupiter; gas giants with little or no solid core. These almost certainly are without moons too. There will be no carbon, and very little chemistry to generate life anywhere in the region.

Population II stars form from clouds of gas and dust that have at least partly been inside another star at some point in the past. When a super-giant star explodes, heavier materials such as Carbon, Oxygen, Magnesium and all the metals including Iron and even Uranium are mixed in with the new star. More importantly, such materials have the potential of producing solid and stable planets such as the Earth and Mars. Many moons will form around those too in time. The seeds are now sewn for life to appear on one or more of those bodies.

There is a possibility that Population III stars exist. These will be very rich in heavy materials, perfect for life to appear as in Population II. None has been confirmed yet and will be rare. If it turns out they don't exist, then it does not cause a problem with our quest as the Sun is a Population II star itself.

All stars live for varying amounts of time. Billions of years of stable output by the mother star is required for life to get started and evolve into more complex forms. The more massive stars such as Betelgeuse in Orion will only live for around 10 million years at most. Not enough time for planets even to form properly then bang! The star explodes and wipe out all chances of life forming there. However, heavy elements essential for life will be ejected ready for chemical and biological processes in the future for new star systems.

A rich star forming region full of materials to generate solid planets with perhaps water, carbon and other chemical compounds to produce life. NASA image

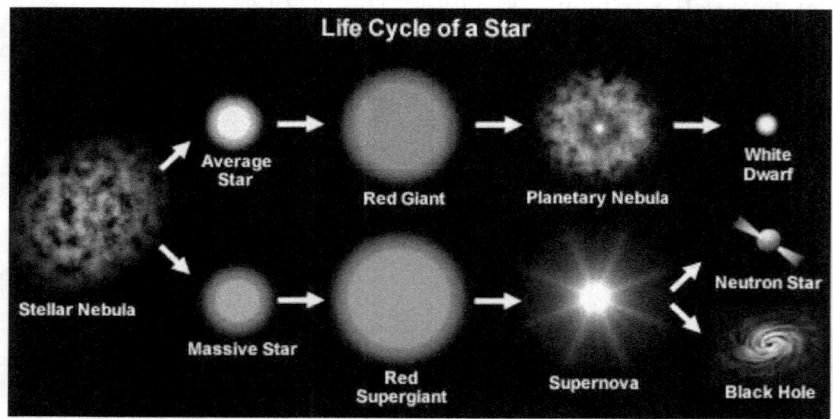

The stars that form with a reasonably low mass evolve gradually as shown on the top half of the diagram. These can live for many billions of years; for most of that time, the star will give out a very steady output of energy. Stars with greater mass at birth have a much shorter lifespan. These will not allow evolution to take place before they explode as a supernova and then become either a neutron star or a black hole.

Therefore, we need to concentrate on Population II stars and then we need to examine their longevity. Stars with a mass of around 40% to 200% of the Sun's mass will exist in a stable form for 4-20 billion years or so. This will allow plenty of time for a species such as ours to arise and perhaps develop the means of communication or even travel between the stars.

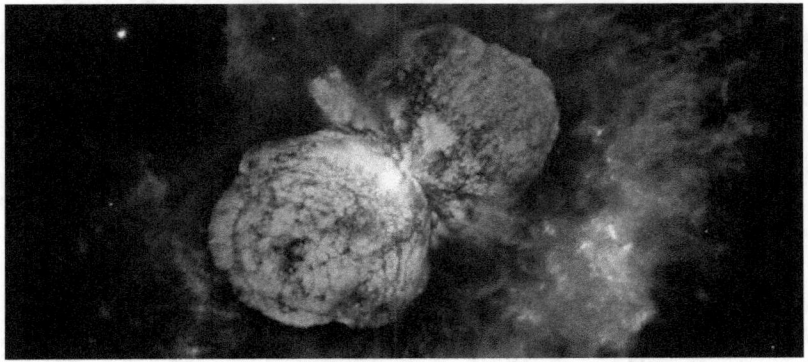

A star about to explode as a supernova. It will generate temperatures of over 1,000,000,000°C for a few minutes. This will create new materials essential for life in a planetary system to be created in the future. NASA image

The sun's period of stability began around 4 billion years ago and will last for another 1 billion years only. Then it will begin to turn into a red giant. The core will increase in temperature enough to push out its surface to perhaps five times its current diameter. This will be enough to boil away the earth's oceans. However, it is not the end for Earth it will occur around 3 billion years later as the Sun swells into a full red giant and almost reach the earth's orbit.

As far as temperature on the planet is concerned then obviously a hotter star will give ideal temperatures further out and a cooler star closer in. As far as we are aware, liquid water has to exist in vast amounts to allow the chemistry for life to develop so the surface temperature needs to fall within a range to allow this.

The green area is known as the 'Goldilocks Zone.' Not too hot, not too cold. Perhaps porridge itself exists somewhere else in the universe and little child aliens are asking amongst each other 'I wonder if there are other planets that is not too hot or cold for mixing porridge.'

There is another flexible angle; if a planet or Moon is too far out from its parent star, then other sources of heat can such as radioactive decay, volcanism, or local gravitational forces can melt ice in large enough quantities to create some forms of life. Several places in our own solar system are perfect examples; the moons Europa, Callisto, Enceladus. It is highly unlikely though that those places will ever produce intelligent civilisations. Finding life of any kind would be a massive leap forward.

A high proportion of stars that we see in the night sky are not single like the sun, they can be double or even multiple systems. Planets have now been detected around such stars in irregular but stable orbits. However, the weather on such planets would almost be too turbulent to develop life, so those stars will have to be discounted for now to be on the pessimistic side. It was thought that as many as 50% were binary or more systems. With new research into star populations, this has now dropped this figure to around 20%, which is good news for alien hunters.

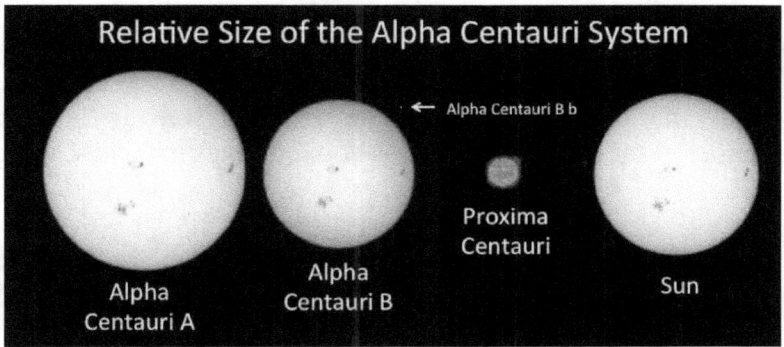

Relative Size of the Alpha Centauri System

← Alpha Centauri B b

Proxima Centauri

Alpha Centauri A

Alpha Centauri B

Sun

Millions of stars similar to the Sun exist in our galaxy alone. Proxima Centauri is almost a red dwarf type star. Although they produce relatively little heat, red dwarfs are very long-lived and incredibly stable, perfect to harbour planets ready for life. We do not need to search for only sun-like stars; this can only serve to widen our hopes.

We now need to move on to define the kind of planets and moons that can harbour life orbiting such suitable stars. The numbers are becoming more accurate by the year as data streams in from observatories on Earth as well as in space. If we keep the numbers real, the end conclusion can be said to be as accurate as we possibly can with current knowledge. We do not need to speculate wildly any longer as similar books have in the past.

Chapter 21   Suitable Planets

Planets are non-luminous spheres with reasonably strong gravity that orbit stars directly. Moons officially are large objects that orbit planets. The true technical term is 'Natural Satellite'... moons will do for this book; it saves ink and reading time.

Mars: In our own solar system, the Earth off course has definite life plus one civilisation with technology. Then there is Mars that could have had life in the past and a small chance some is still thriving today. Once life has a hold, many species evolve to suit various latitudes, depths of water, underground in caves, soil and even inside solid rock hundreds of metres down.

As the climate changed to what we see today, there is a chance that some species adapted and continued to live quite happily underground with complete disregard as to how the climate changed on the surface. This scenario would almost certainly occur on Earth. In around 1 billion years from now, the Sun will begin to expand due to increasing density of the core. As the surface area increases, the temperature on the Earth will rise and boil away the oceans and atmosphere. The surface will become sterile but life forms living deep underground will be unaffected.

Recent Positive News: It has been discovered that Methane Gas is being produced from the surface. It belches out clouds of it, a gas that on Earth mostly comes from living organisms. When animals and other organisms eat food, they produce methane as a waste gas. From one end or the other, that gas finds its way out into the air.

The Curiosity Mars Rover has detected Methane directly from the surface. Microbes could be living under the Martian surface and churning out the gas. But any number of other processes may be responsible instead. Detecting methane is never enough to answer the question of whether or not we are alone.

Scientists made its first detection of methane on Mars in 2009. Telescopes picked out spectral features of the gas, and revealed plumes of methane pouring from the northern hemisphere. The finding was evidence for a replenishing supply of the gas on the planet, because methane molecules, on average, last only 340 years in Martian air before they are broken down by sunlight and Oxygen.

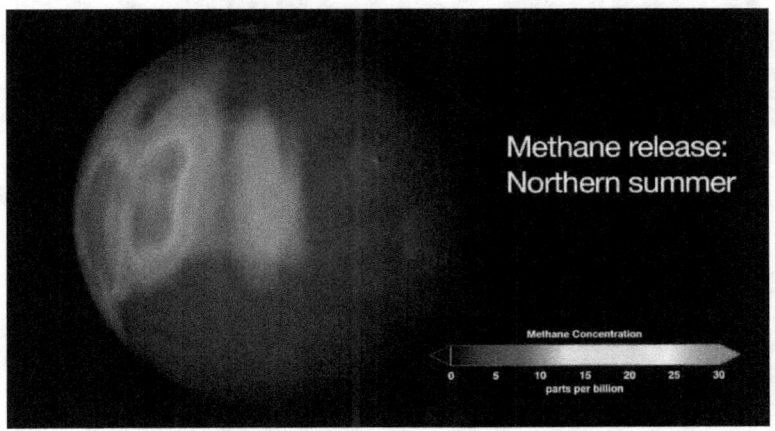

Methane emissions as mapped by a NASA Orbiter. As of October 2016, the European Space Agency now has a dedicated satellite mapping the precise locations where the Methane is coming from.

The original atmosphere on Mars was much denser and warmer due to volcanic eruptions. After the core cooled, the eruptions ceased and the atmosphere was slowly lost until just 7 millibars of pressure is left today compared to the earth's 1000 millibars.

It is possible that the Methane source is related to a volcanic process instead. Volcanoes and other similar forms of out-gassing from the Earth create Methane, but such mechanisms have no connection with the weather, so why is the Methane emissions on Mars seasonal? There is currently only one answer; the origin is biological.

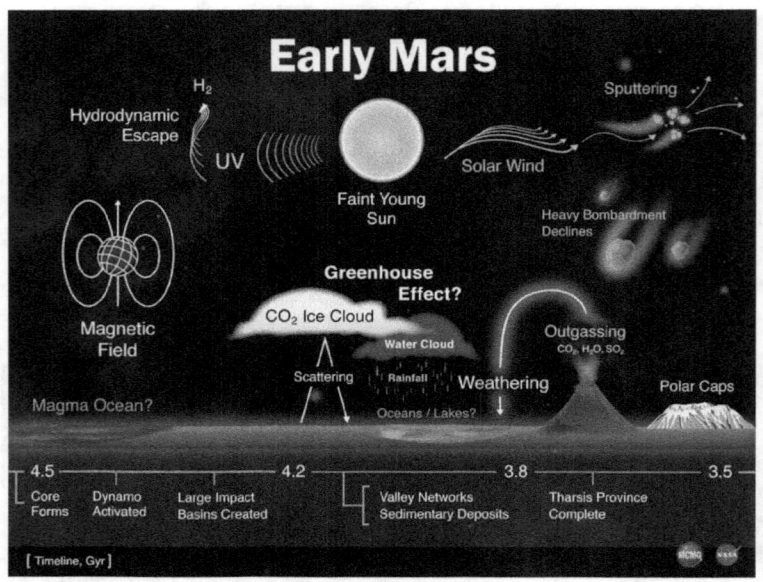

Above; Mars as it may have performed 3 billion years ago.
Topographical studies from Orbiting spacecraft have mapped the
low-lying areas and Landers have confirmed sediment deposits in
such regions. One large ocean certainly existed on Mars allowing
the right chemistry to produce life.

Three more spacecraft are being built to discover current life
on Mars and were all launched in 2020. The European rover
called Exo-Mars partly made in Stevenage UK (above), one
by NASA and the other by China.

Venus: Venus possibly had oceans on it too in its early history but any evidence of life that far back in time would now have been destroyed due to extreme volcanic processes that cover the entire surface. So, there is one definite planet and two possible planets out of eight that is, or perhaps was, a home for life in general. Out of those, one produced a race that can build rockets, produce beer and write books.

Perhaps life got a head start on Venus then was wiped out completely by the later extreme processes that we see today. A European spacecraft known as Venus Express, revealed numerous volcanic features, far more than on Earth. The very young global surface with very few meteor impacts is further evidence of that. Any fossilised life forms in this early epoch would now have long been obliterated.

Venus as it may have looked in its early history. It is closer to the Sun than earth, and had a larger core which must have resulted in extra volcanism. Enormous amounts of Carbon Dioxide would have been released and created a runaway greenhouse effect.

The Venus Express data showed that Venus almost certainly was a watery world in its early history.

At present, only the planet Earth definitely harbours life in our own solar system. However, with two possible other planets just either side us showing evidence of similar habitats in the distant past, expands the possibilities of life occurring elsewhere enormously. At the same time, without the solid evidence we need, the Earth could still be the only one in the entire Universe that is the home to intelligent creatures or even living organisms in general.

Beyond the Solar System: We now should begin to investigate the larger picture. So far, we have only looked around our own back yard, around one single star; the Sun. We have another 200 billon stars to consider in the Milky Way galaxy alone, and 14 million galaxies have been observed directly so far with billions more to find. This book is about detecting civilisations directly from Earth, so to be practical, we will only concern ourselves with our own galaxy from this point.

A great science fiction novel called 'Tau Zero' by Poul Anderson, (published in 1970) was about a spacecraft crewed by 25 men and 25 women aiming to reach a distant star system 33 light years away. As they got closer, they crew realised the planets there not suitable, so they headed off for another that also turned out to be too harsh for colonisation. The crew looked at each star system upon approach through a powerful telescope, as there was no other method to detect and study such planets. The story is based on technologies that may exist in the late 22nd Century. Here we are in the early 21st Century and we have methods of detecting earth-like planets many times further out and in detail than mentioned in Tau Zero.

The rapid acceleration of our knowledge of the universe shows no sign of slowing either. Some scenarios seem to demonstrate we are falling behind as with the movie 2001 A Space Odyssey. According to that story, we should not only have a moon-base, but also a manned ship heading for Jupiter with a crew partly in suspended animation. That has not even occurred upon writing this book, but the 'Skyping' prediction did, digital cameras and a space station; but non-rotating.

In some ways, science fact has overtaken science fiction. We have methods of detecting planets orbiting other stars hundreds of light years away and we are just on the verge of even studying the atmospheres of some of those planets.

The Kepler satellite has often hit the news with discovering most of the Extra-Solar Planets we now know of. A technique has been developed that is related to Solar Eclipses – the Transit Method. Such techniques are discussed later.

The new James Web Telescope, launch on 25th December 2021 was designed to gather data on planetary atmospheres outside our solar system. If an Earth type planet is found to have an Oxygen / Nitrogen rich atmosphere, then we can reasonably safely conclude that life on that planet may be generating the Oxygen through photosynthesis as on earth.

The James Web Telescope was placed in orbit
1.5 million miles from Earth in 2021. ESA image.

The telescope is almost a global effort as was built and funded by the NASA, European Space Agency and the Canadian Space Agency; a technical effort rarely seen in history.

As of 2016, it has been confirmed that a planet with a similar mass to the Earth has been discovered orbiting Alpha Centauri B and is well within the habitable zone. Along with all the discoveries made so far, we now have a fairly accurate figure of how many stars have planets, and we are beginning to develop a maximum and minimum range of how many earth-type planets exist around sun-like stars too. Such figures will be revealed in the 'Final Figures' chapter. Yes, a little maths; sorry. Only multiplication and fractions will be used so relax.

Above; an artist's impression of the nearest planetary system
to earth, just 4.2 light years away.

Another recent discovery involves rogue planets. These are also known as an interstellar planet / nomad planet / free-floating planet / orphan planet / wandering planet or starless planet. They orbit the Milky Way or any other galaxy directly but without a parent star. Such objects have either been ejected from the star system in which they formed by a close encounter by another planet or perhaps never been bound gravitationally to any star in the first place.

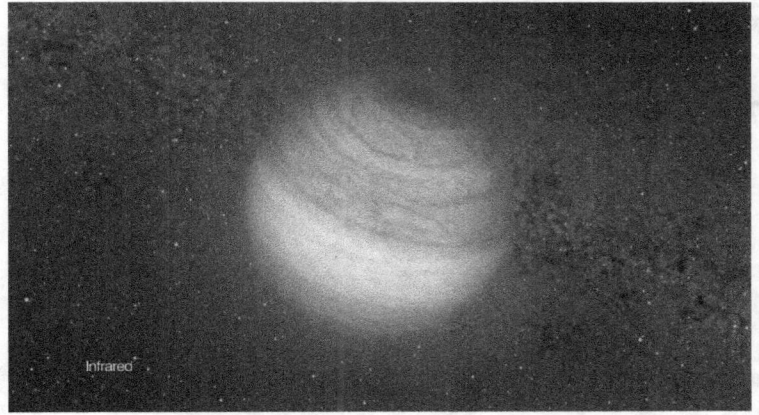

Rogue planets will be unlit by a parent star but may be visible in the infrared as they can produce heat internally. Some rogue planets will be visible in the infrared. As of 2016, four so far have been confirmed. (Artist Impression – NASA)

There may be billions of rogue planets in the Milky Way, some are thought to have formed in a similar way to stars, and the International Astronomical Union (IAU) has proposed that those objects be called sub brown dwarfs. Possible examples have recently been found, one is Cha 1109913 – 773444; a great name. The closest free-floating planetary-mass object to Earth yet discovered is seven light years away, WISE 0855-0714.

Planets such as these do not have an external source of heat to generate life; it is possible they have in internal source instead in a similar fashion to Jupiter or even the earth. The source can be either trapped radioactivity in the core or simply due to gravitational pressure. The heat will radiate out to the surface and give off an infrared image of itself to the universe. Such planets should not be ruled out entirely for harbouring life of some kind but, as they are incredible difficult to gain any information on them due to their dark environment, then we may discount such objects in our statistics for now.

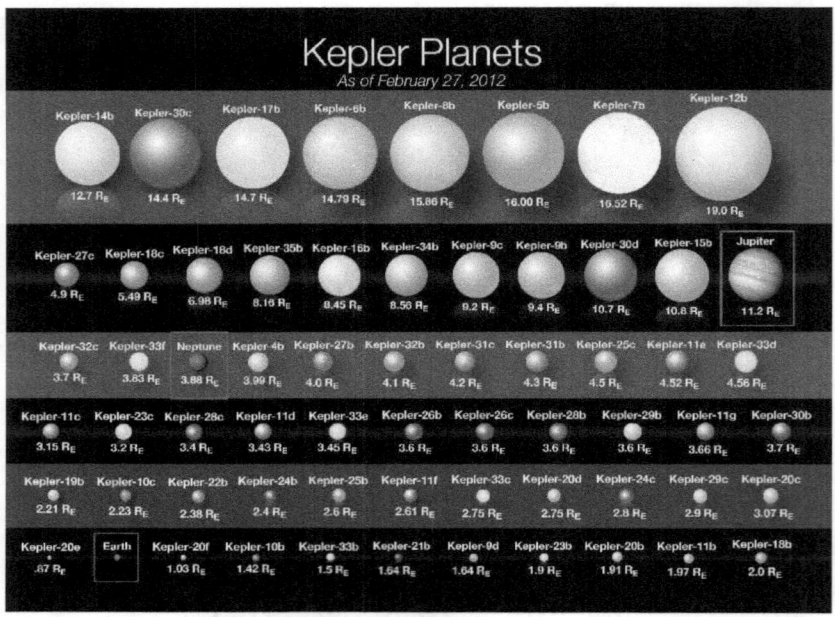

Size comparison of Exoplanets discovered by the
Kepler Space Telescope. Credit ESA.

Chapter 22   Suitable Moons

Thanks to movies such as Star Wars, our minds open to the possibility that life can develop on moons of planets, not just planets themselves. As moons can out number planets in almost every planetary system, this pushes up the numbers of potentially life bearing bodies.

No moons have yet been confirmed outside the solar system but we can assume they do indeed exist. As almost 200 have been discovered and imaged close up, we can put forward examples of places that could harbour life forms. They are presented here in order from the sun.

The Moon: This is the first Moon from the sun. It orbits the Earth every 29 days and 13 hrs on average. We have known for centuries it has no atmosphere, no known active volcanism; it is essentially a dead world but not uninteresting from a geologist's viewpoint.

Many UFO supporters believe the Moon is inhabited by Alien beings today. They claim to have seen buildings, flying craft, runways etc across the Moon by studying images from the probes. One claims to have discovered a 1000-mile-long wall; nope! It just a missing line of data. If it was that long, we could see it through binoculars from our back gardens. Quietly ignore this hypothesis.

If alien beings have used the Moon as a base to study earth, it would make a little sense but such evidence is just not there. The Moon has been mapped to a resolution of around 50cm, so anything larger than that would have shown. The airless environment there is perfect for preserving such evidence for millions of years, but nothing has ever been seen.

There is no cover up either. NASA has often been accused of keeping vital discoveries secret. How can this be when Europe,

Japan, China, Russia have sent mapping missions to the Moon too? NASA cannot control data from other nations.

If alien buildings have been found, such a discovery would spur on fantastic manned missions to investigate such features. Conspiracy supporters will continue to produce ridiculous arguments of blurred images that 'look like' buildings and such to a non-lunar-geologist. There are plenty of features on Earth that look like pyramids when the Sun is at a particular angle. These are clearly mountains instead; just ask the mountaineers. Sand dunes can be imagined as pyramids as the wind direction changes; all quite normal.

Europa: This is probably the most promising Moon to be the home to aquatic life forms. Under the surface of thick ice there should exist a global deep ocean. So much water in fact that it probably exceeds the earth's supply in volume.

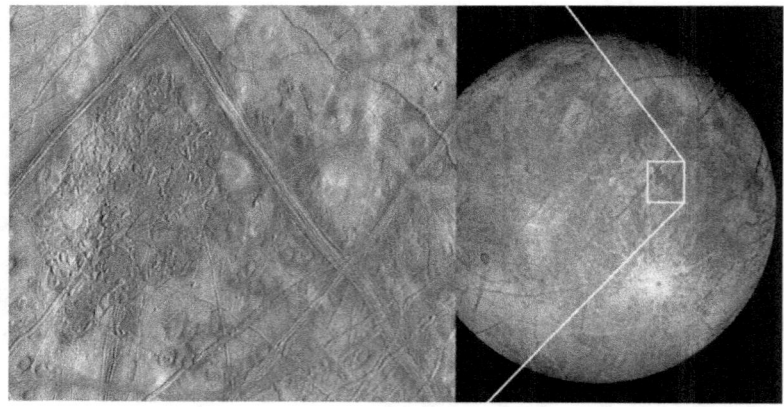

From models of magnetic forces, and images of its surface, scientists have long suspected that a giant ocean, roughly 160km (100 miles) deep, lies somewhere between 10-30km beneath the ice crust.

Jupiter's gravity tugs on Europa in an uneven way as it orbits and rotates; this produces a flexing on its diameter and generates heat. This is known as gravitational tidal heating and can be energetic enough to melt the subsurface ice. Complex forms of chemistry may well take place within the melted ice to produce life. However, as biologists still cannot agree as to how this occurs, the exacting conditions still cannot be defined.

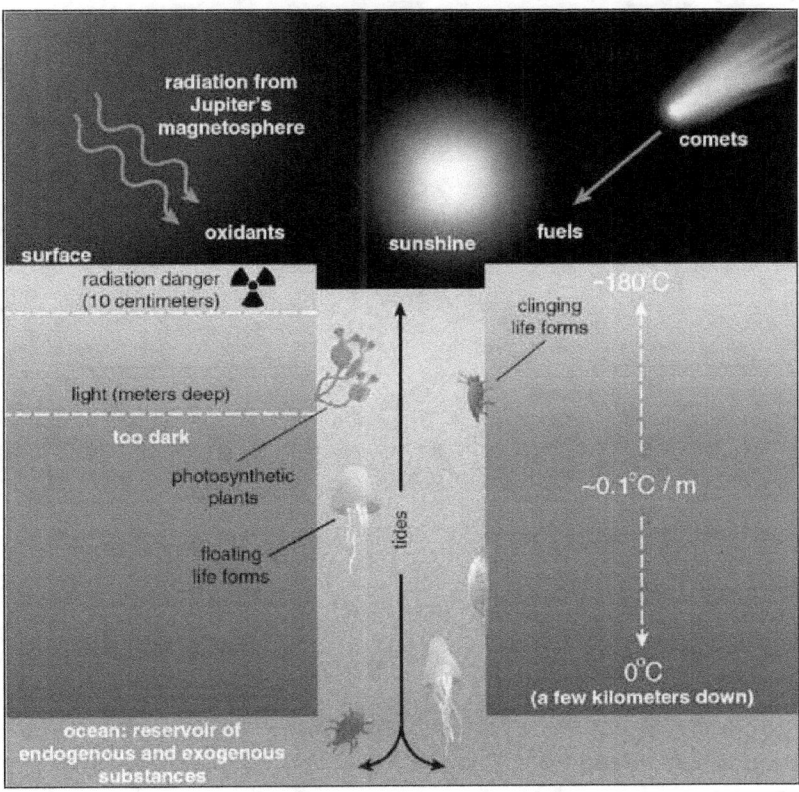

For evolution to take place, a very stable environment with a constant source of heat is required. The suspected ocean below Europa's ice would certainly have these properties. If the liquid water on the following bodies were drained and placed into a sphere, the following diagram shows clearly home much water they contain compared to each other.

Liquid Water in the Solar System

EUROPA

TITAN

EARTH

CREDIT: PHL @ UPR Arecibo, NASA

Out of these three examples, the Earth has the least amount of water in volume and yet is teaming with life. Water is obviously not the only ingredient required to produce living organisms, but is essential as far as we are aware.

Callisto: Much like Europa and Ganymede, and Saturn's moons of Enceladus, Mimas, Titan and Dione, the possible existence of a subsurface ocean on Callisto has led many scientists to speculate about the possibility of life. This is particularly likely if the interior ocean is made up of salt-water, since halophiles (which live in high salt concentrations) could in theory live there.

In addition, the possibility of extra-terrestrial microbial life has also been raised regarding Callisto. The environmental conditions necessary for life to appear are more likely on Europa and Ganymede. The main difference is the lack of contact between the rocky material and the interior ocean, as well as the lower heat flux in Callisto's interior.

While Callisto possesses the necessary pre-biotic chemistry to host life, it almost certainly lacks the necessary heat energy.

Therefore, the most likely candidate for the existence of extra-terrestrial life in Jupiter's system of moons remains Europa.

Titan: This is the Saturn's biggest Moon and if fact the largest in the solar system. It is often described as a planet-like Moon because it is 50% bigger than our Moon and 80% more massive. The most exciting feature of Titan is that it has methane and ethane clouds.

These clouds have formed lakes, rivers and seas of methane and ethane, and the Moon has a climate including wind and rain.

It even has seasonal weather patterns like earth. Titan's methane cycle is much like earth's water cycle, but at a much lower temperature. Worlds like Titan are too cold for liquid water to exist anywhere on the surface.

The extreme cold of the Moon puts liquid water out of reach, and it is buried 31-62 miles (50–100 km) below a frigid ice crust. Some scientists are exploring the possibility of whether a different kind of life could come about on these types of places.

In a paper published in the journal PNAS, a group of scientists modelled whether a certain type of life-giving reactions might be possible under the conditions present on Titan. Their simulations found a polymer, called polyimine, which forms from hydrogen cyanide (HCN), could drive chemistry on the surface of the moon.

The European Huygens probe touching down on Titan 2005

A study in November 2014 argued that alien species could exist on planets that contain an exotic substance known as 'supercritical' carbon dioxide, rather than water. This type of CO_2 is created when liquids and gases reach their temperature and pressure thresholds, creating a supercritical fluid that has features of both a liquid and gas.

Carbon dioxide becomes supercritical when its temperature exceeds 32°C and its pressure goes beyond the standard atmosphere at sea level. On Earth, it is increasingly used in applications such as dry cleaning or to sterilise medical equipment, but astro-biologists at Washington State University believe it could also be capable of sustaining life without any water.

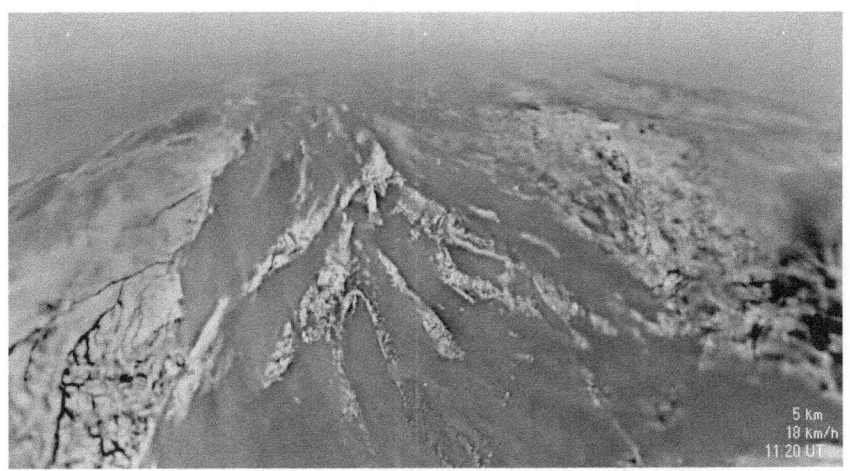

Actual image of a mountain on Titan during the descent.

Enceladus: Enceladus is another member of a family of icy worlds populating our outer solar system. These bodies are some of the most promising places for life because they receive tidal heat energy from the gas giants they orbit through gravity and have substantial amounts of liquid water.

The Cassini spacecraft has been taking regular measurements of Enceladus for more than a decade to evaluate its environment. One of the key factors influencing the habitability of an environment is its chemical composition, in particular its pH or Acidic Level. On Earth, it is possible for life to exist in a wide range of the pH scale from 0 (battery acid) to 14 (drain cleaner). Knowing the pH can help us to identify reactions that affect the probability of an environment to contain living organisms.

We cannot stick a strip of pH paper into the ocean on Enceladus and test it, but can be estimated by looking at molecules in the plumes as they gush out into space.

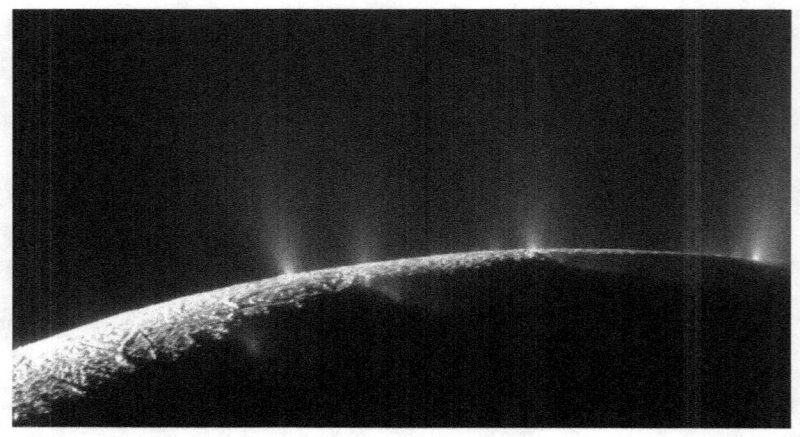

The plumes of Enceladus that have been analysed by the Cassini spacecraft. The Hubble Space Telescope from Europa has observed similar outbursts as of 2016.

Recently, geochemist Christopher Glein led a team that developed a new approach to estimating the pH of Enceladus' ocean using observational data from the Cassini spacecraft. Scientists can now analyse water on the moons of Saturn thanks to measurements by the Cosmic Dust Analyser (CDA) and carbon dioxide gas by instruments on board Cassini.

Glein's team tried to create the most accurate chemical model to date of the ocean by using data from two instruments. Their model suggests that Enceladus has a sodium, chloride and carbonate ocean with an alkaline pH of 11 or 12, close to the equivalent of soapy water. The estimated pH is slightly higher by 1 to 2 units than an earlier estimate based on CDA data alone, but the different approaches are consistent in terms of the chemistry of an ocean that is suitable for life.

The Cassini Mission that accumulated this data ended on 15 September 2017. The probe was deliberately aimed at Saturn's atmosphere and successfully burnt up. The scientists didn't want the risk of contaminating Enceladus by accidentally colliding with it in the future.

Suitable Asteroids: We have discussed Planets and moons that could harbour life, but what about asteroids? These are very small bodies up to 900km across orbiting the Sun between Mars & Jupiter. None have an atmosphere so the chemistry on them should be far too basic to ever create life. Scientists searching for evidence of life beyond Earth have now discovered organic material on asteroid Ceres, this is also now classed as a dwarf planet.

The carbon-based materials, similar to what may have been the building blocks for life on Earth, were discovered by NASA's Dawn space probe in February 2017. The exact molecular compounds in the organics cannot as yet be identified, but they match tar-like minerals such as kerite or asphaltite.

"The discovery indicates that the starting material in the solar system contained the essential elements, or the building blocks, for life," said Dawn's lead scientist, Christopher Russell.

"Ceres may have been able to take this process only so far. Perhaps to move further along the path took a larger body with more complex structure and dynamics (like Earth)."

Chapter 23 Exoplanet Discoveries

For centuries, people of all ages have looked out at the night sky and wondered if they were alone in the universe. With the discovery of the planets Uranus and Neptune in 1781 and 1846 respectively, then the true size of the Milky Way galaxy became known and other galaxies beyond ours, this question has only deepened and became more important and profound.

Astronomers have long suspected that other star systems in our galaxy and beyond had planets of their own. It has only been within the last few years that any have been observed directly at all. The various methods employed for detecting these extra-solar planets / exoplanets have improved, and the official list as of June 2023, reach over 5,200.

An extrasolar planet, also known as an exoplanet, is a planet that orbits a star other than our own. Our Solar System is only one among billions in the Milky Way alone. Other galaxies will be the same but we will concentrate on ours, as the others are far too distant to observe in any detail. The 200 billion stars in the Milky are enough to contend with.

In the 1500s, there have been astronomers who hypothesized of the existence of exoplanets. Italian philosopher Giordano Bruno, an early supporter of the Copernican theory that planets orbited the Sun rather than the earth, made the first known mention. He also put forward the view that the stars are similar to the Sun and may well be accompanied by planets.

In the 1700's, Sir Isaac Newton made a similar suggestion in his famous book on Gravity – Principia. He wrote *"And if the fixed stars are the centres of similar systems, they will all be constructed according to a similar design and subject to the dominion of one."*

Since Newton's time, various claims have been made, but which were rejected by the scientific community as false positives. In the 1980's, some astronomers claimed that they had identified some extrasolar planets around nearby stars, but were unable to confirm their existence.

One of the reasons why exoplanets are so difficult to detect is because they are hundreds of times fainter than the stars they orbit. Additionally, these stars give off light that obscures them from direct observation. As a result, the first true extrasolar planet discovery was not made until 1992 when astronomers Aleksander Wolszczan and Dale Frail, using the Arecibo Radio Observatory in Puerto Rico, observed several earth-mass planets orbiting the neutron star PSR B1257+12.

A neutron star (sometimes referred to as a Pulsar) has an enormous gravitational pull. Many years ago, it would have been a red super-giant star that exploded (supernova). Such stars are very short lived and little time would be available for life to evolve on planets orbiting them. If any planet was fortunate enough to develop life, it would have been sterilised within hours of the supernova event. The core of the star then collapses and become incredibly dense, hence creating a strong gravitational field.

It was not until 1995 that the first confirmation of an exoplanet orbiting a normal main-sequence star was observed. In this case, the planet observed was 51 Pegasi b, a giant planet found in a four-day orbit around the Sun-like star 51 light years away.

Initially, most of the planets detected in the early years were gas giants larger than Jupiter; sometimes referred to as Super Jupiter's. Far from suggesting that gas giants were more common than rocky earth-like planets, these discoveries were simply because massive planets are simply easier to detect.

Range of Detection Methods: While some exoplanets have been observed directly with telescopes (a process known as Direct Imaging), the vast majority have been detected through indirect methods such as the transit method and the radial-velocity method; observing a star's wobble caused by the gravity of a planet. (See the supporting website for animations on how this works).

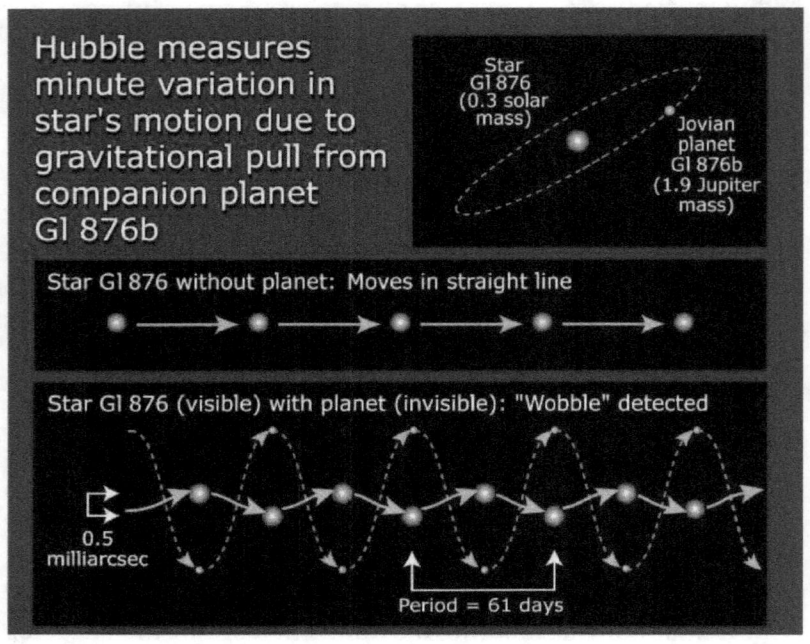

By observing a star wobbling around, proves the existence of an unseen orbiting body; a planet. The timing of the wobble equals the planet's year; the distance of the planet from the star can be deduced as well as an indication of its surface temperature. The amount of wobble shift denotes the planet's mass and hence surface gravity.

The Kepler Mission: Named after the revolutionary astronomer Johannes Kepler, the Kepler Space Observatory was launched by NASA on 7 March 2009 for discovering Earth type planets orbiting other stars. As part of NASA's Discovery program, a series of relatively low-cost satellites focused on scientific research, Kepler's mission is to survey a portion of our region of the Milky Way and find evidence of exoplanets. This is to estimate how many stars in the Milky Way have planetary systems like ours; a vital component of the Drake Equation.

Relying on the Transit Method of detection, Kepler's sole instrument is a photometer instrument that continually monitors the brightness of over 145,000 stars in a fixed area of the sky in Cygnus the Swan. This data is transmitted back to Earth where it is analysed by scientists to look for any signs of regular dimming caused by extrasolar planets passing in front of their host star.

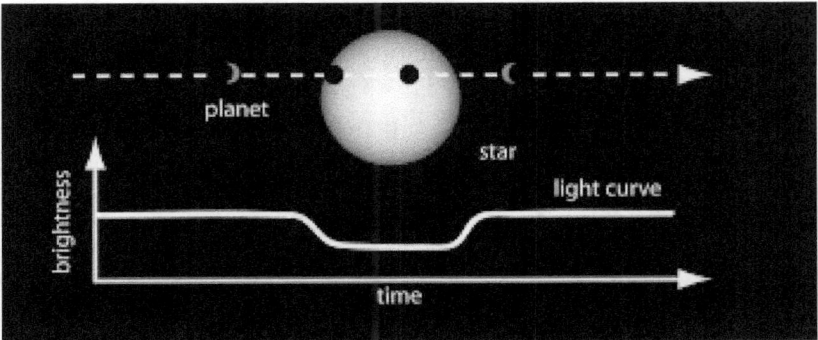

The initial planned lifetime of the Kepler mission was 3.5 years, but greater-than-expected results led to the mission being extended. In 2012, the mission was expected to last until 2016, but this changed due to the failure of one the spacecraft's reaction wheels – these are used for pointing the spacecraft on a target area of the sky. On 11 May 2013, a second of four reaction wheels failed, and threatened the entire mission.

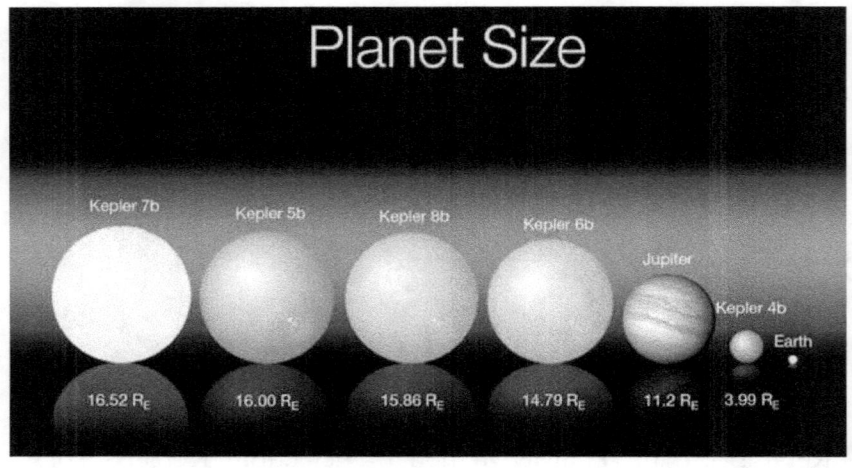

Planet Size

Kepler 7b	Kepler 5b	Kepler 8b	Kepler 6b	Jupiter	Kepler 4b	Earth
16.52 R$_E$	16.00 R$_E$	15.86 R$_E$	14.79 R$_E$	11.2 R$_E$	3.99 R$_E$	

Examples of giant Jupiter's by size comparison discovered by Kepler. These were the easiest to find, as the dip in brightness of a star would be large as it passed in front of it.

On 15 August 2013, NASA admitted that they had given up trying to fix the two failed reaction wheels and modified the mission instead. Rather than scrap Kepler, it was proposed changing the mission to detect habitable planets around smaller and fainter red dwarf stars. This became known as the K2 Second Light mission for the probe, and the budget was approved on 16 May 2014.

Since that time, the K2 mission performed better than expected has focused more on brighter stars. As of December 2017, astronomers have confirmed the presence of 5,944 exoplanets that include 748 multi-planet systems. The majority of which were found using data from Kepler. The probe observed over 150,000 stars in the course of its K1 and K2 missions.

By exploring the data, the radius, mass and distance from the parent star can be established. From February 2020, the radii of 3,162 has been determined, and the mass for 904 of those.

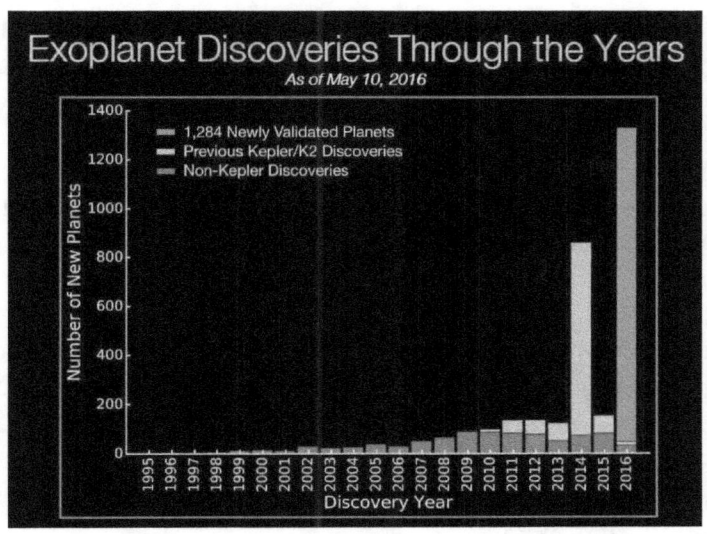

This graph clearly shows the exponential rate of planet discoveries. One planet was found on 1995, 1284 planets in just over four months of 2016. NASA

In November 2013, astronomers reported that there could be as many as 40 billion Earth-sized planets orbiting in the habitable zones of life – friendly stars within the Milky Way. It is estimated that 11 billion of these planets may be orbiting Sun-like stars. These stars are stable enough to produce life on its orbiting planets, and then we add on statistics for moons too. In our solar system, there are realistically four moons that just could be a home to living organisms. The final figure for the Milky Way comes out to 200 billion potential homes for life; an essential figure for the Frank Drake Equation.

Habitable Earth Type Planets: The discovery of exoplanets has also intensified interest in the search for extra-terrestrial life, particularly for those that orbit in the host star's habitable zone. Also known as the goldilocks zone, this is the region of the solar system where conditions are warm enough so that it is possible for liquid water and perhaps life to exist on the planet's surface.

The first planet confirmed by Kepler to have an average orbital distance within its star's habitable zone was Kepler 22b. This planet is located about 600 light years away in the constellation of Cygnus, and was first observed on 12 May 2009, and then confirmed in Dec 2011. Based on all the data obtained, scientists believe that this world is roughly 5 times the diameter of Earth, and is likely covered in oceans or has a liquid or gaseous outer shell.

Before the launch of Kepler, the vast majority of confirmed exoplanets fell into the category of Jupiter-sized or larger. However, as of September 2019, Kepler has identified 55 Earth-size or "Super-Earth" size planets that are also located in the habitable zone of their parent stars, and some of those around Sun-like stars.

Current Potentially Habitable Exoplanets
Ranked in Order of Similarity to Earth

#1	#2	#3	#4	#5	#6
Kepler-62 e 0.83	Gliese 667C c 0.82	Gliese 581 g* 0.82	Tau Ceti e* 0.77	Gliese 667C f 0.76	Kepler-22 b 0.75

Earth 1.00
Mars 0.64

#7	#8	#9	#10	#11	#12
Gliese 163 c 0.74	HD 40307 g* 0.72	Kepler-61 b 0.72	Kepler-62 f 0.67	Gliese 667C e 0.60	Gliese 581 d 0.53

Neptune 0.28
Jupiter 0.16

*planet candidates   Number below the names is the Earth Similarity Index (ESI)   CREDIT: PHL @ UPR Arecibo (phl.upr.edu) June 26, 2013

According to a recent study from NASA Ames Research Center, analysis of the mission data indicated that about 24% of M-class stars might host potentially habitable, Earth-size planets or no more than 3 times the earth's diameter. Based upon the number of such stars in the galaxy, that alone represents about 10 billion potentially habitable, Earth-like worlds.

Meanwhile, the K2 phase suggests that about one-quarter of the larger stars surveyed may also have Earth-size planet orbiting within their habitable zones. Taken together, the stars observed by Kepler make up about 70% of those found within the Milky Way. So one can estimate that there are literally tens of billions of potentially habitable planets in our galaxy alone. This singular discovery gives us an indication of how many planets may indeed host living organisms of some kind.

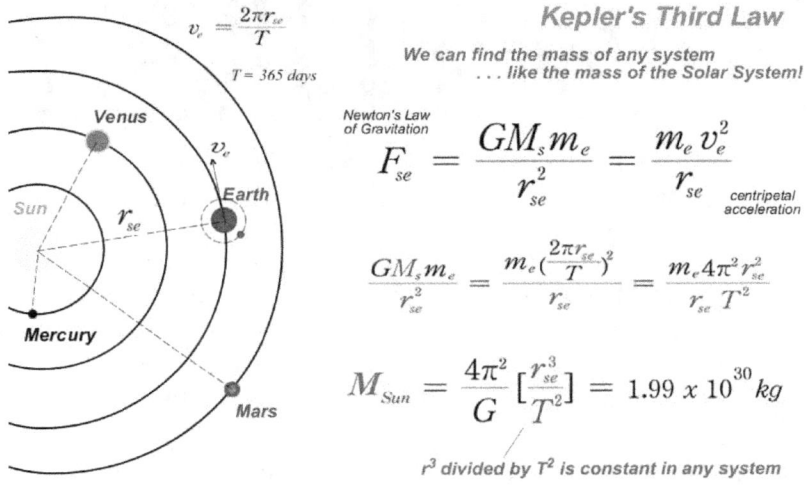

Kepler's Third Law

We can find the mass of any system
. . . like the mass of the Solar System!

Newton's Law of Gravitation

$$F_{se} = \frac{GM_s m_e}{r_{se}^2} = \frac{m_e v_e^2}{r_{se}} \text{ centripetal acceleration}$$

$$\frac{GM_s m_e}{r_{se}^2} = \frac{m_e (\frac{2\pi r_{se}}{T})^2}{r_{se}} = \frac{m_e 4\pi^2 r_{se}^2}{r_{se} T^2}$$

$$M_{Sun} = \frac{4\pi^2}{G} [\frac{r_{se}^3}{T^2}] = 1.99 \times 10^{30} kg$$

r^3 divided by T^2 is constant in any system

Diagram labels: $v_e = \frac{2\pi r_{se}}{T}$; $T = 365\ days$; Venus; v_e; Earth; Sun; r_{se}; Mercury; Mars

The formula derived by Johannes Kepler in 1619 allow us to calculate several factors such as Radius & Mass of orbiting planets and binary star systems. This removed estimates in one single formula.

Atmosphere Detection: In the case of the Transit Method, a planet is observed when crossing the path (transiting) in front of its parent star's disk. When this occurs, the observed brightness of the star drops by a small amount. This can be measured to determine the size of the planet, and it has the benefit that it sometimes allows a planet's atmosphere to be investigated through spectroscopy.

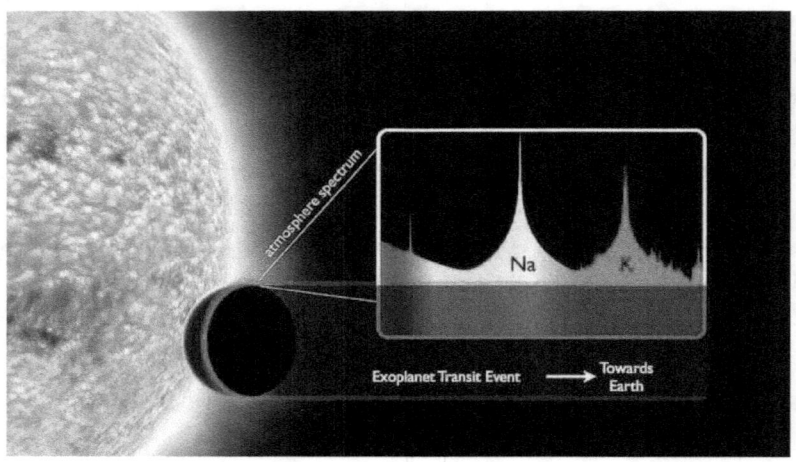

By passing light through as instrument called a Spectroscope, we can deduce what a star or an atmosphere of a planet is comprised of without travelling to it.

However, such methods also suffer from a high rate of false alarms, and generally require that part of the planet's orbit intersect a line-of-sight between the host star and Earth. As a result, confirmation from another method is usually considered necessary or more importantly, requires repeat observations. Nevertheless, it remains the most widely-used means of detection and is responsible for the vast majority of all exoplanet discoveries The Kepler telescope uses this method.

In 2001, the Hubble Space Telescope made the first direct detection of the atmosphere of a planet orbiting a star outside our solar system. This observation obtained the first information about its chemical composition. This demonstrated that it is possible with Hubble and other telescopes to measure the chemical makeup of extrasolar planet atmospheres and to potentially search for chemical markers of life beyond Earth.

The planet orbits a yellow, Sun-like star called HD 209458, a seventh magnitude star (visible in binoculars), which lies 150 light-years away in the autumn constellation Pegasus. Its atmosphere was probed when the planet passed in front of its parent star, allowing astronomers for the first time ever to see light from the star filtered through the planet's atmosphere.

Lead investigator David Charbonneau of the California Institute of Technology (Pasadena, California) and colleagues used Hubble's spectrometer to detect the presence of sodium in the planet's atmosphere.

"This opens up an exciting new phase of extrasolar planet exploration, where we can begin to compare and contrast the atmospheres of planets around other stars."

The astronomers actually saw less sodium than predicted for this Jupiter-class planet, leading to one interpretation that high-altitude clouds in the alien atmosphere may have blocked some of the light.

The Hubble observation was not tuned to look for gases expected in a life-sustaining atmosphere, but does demonstrate this unique observing technique opens a new phase of studying extrasolar planets. Such observations could potentially provide the first direct evidence for life beyond Earth by measuring unusual abundances of atmospheric gases caused by the presence of living organisms.

The planet discovered in 1999 through its slight gravitational tug on the star. Based on that observation the planet is estimated to be 70 percent the mass of the giant planet Jupiter (or 220 times more massive than Earth). It is an ideal target for repeat observations because it transits the star every 3.5 days - which is the extremely short amount of time it takes the planet to whirl around the star at a distance of merely 4 million miles from the star's searing surface. This precariously close distance to the star heats the planet's atmosphere to 1100° Celsius.

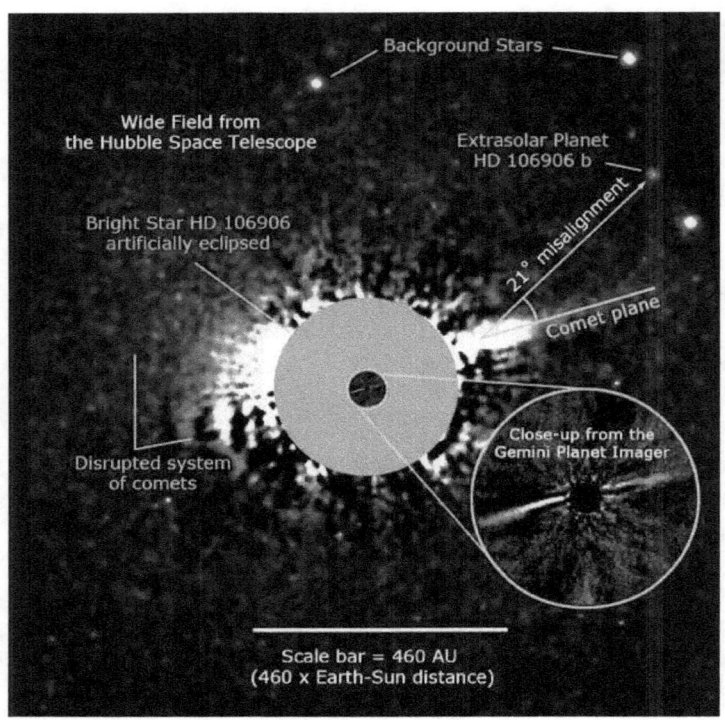

Planets have been directly photographed orbiting the star Beta Pictoris by the Hubble Space Telescope. NASA

Related Future Missions: With the winding down of Kepler's mission, and so many discoveries made within a short period, NASA and other federal space agencies plan to continue in the hunt for extrasolar planets. NASA missions that has picked up where Kepler has left off include the Transiting Exoplanet Survey Satellite (TESS), launched on 18 April 2018, plus the James Web Space Telescope, launched on 25th December 2021.

The European Space Agency (ESA) hopes to continue to map out a significant portion of the Milky Way Galaxy using its Gaia satellite; this commenced operations in 2013. The Herschel Space Observatory (ESA / NASA), has been in operation since 2009 and is expected to make many more interesting discoveries in the coming years.

Detecting Life: Once an earth-like extrasolar planet was discovered the natural question arose of whether there is a method of detecting the presence of life on it from here. The good news is that there is. By splitting the light of the planet into a spectrum (rainbow), we can determine what gasses make up the atmosphere if it has one. If Nitrogen and Oxygen are found, then we may have a positive answer of yes there may be life. This may sound simple but is incredibly difficult in reality. The James Web Telescope is doing just that.

The Kepler Space Telescope concentrated on a very small part of the Milky Way and yet found hundreds of new planets. This statistical analysis is a large enough sample to give some fairly accurate numbers regarding possible life bearing planets in our galaxy.

Chapter 24   Arising Intelligences

We automatically imagine the human beings are the most intelligent species on this planet. In biology, 'intelligence', in the broadest sense of the term, refers to the ability of a life form to adapt to its environment through learning and then shaping the surroundings to suit itself. 'Intelligence', in that sense, translates as the ability of a creature to show such adaptive ability. There are other aspects too that involve communication skills, understanding the needs across species for their own advantage, deciding which member of a species is friendly and which is not.

Some example: Bees: Communicating through dance. They have dance-offs to take up decisions. They dance to show the direction of the best flower they found. Remembering where the best flowers are after flying miles is fascinating point in itself.

Crows and Ravens: These birds can understand the concept of Archimedes principle. They use this knowledge to change water levels to access food by dropping stones into a deep bowl of water to make it rise for easier access. They are known to multi-task too (unlike most men).

Dogs: They are rapid learners. They can grasp up to 1000 different words.

Hedgehogs; Produce a tour of gardens for food and socialising with other hedgehogs and humans. The route adapts with new discoveries. Apart from this, they are rather dumb.

Elephants: Although memory is not the finest point to define intelligence, it is fascinating to note that elephants have great memories. They can remember friends and/or enemies for decades, depending on their health.

Ants: They can navigate long routes. Their social behaviour is similar to humans. Each ant plays a specialised role within their own community for a share of the food and security. The number of ants playing each role is in a similar proportion with other unrelated nests.

Many biologists agree that Dolphins are probably the most intelligent species set against our own. There are many reasons for this conclusion;

They behave almost similar to humans. They socialize in gangs that contain about 2 - 15 individuals. They show aggressive behaviour to show dominance when needed. Large adult males often roam the periphery of a group, and may afford some protection against predators. They protect the wounded fellows until they are cured

Being playful shows intelligence. Whales and dolphins are one of the few animals that show this behaviour. They enjoy a game of catch, which is evident from the dolphin shows. They create their own games, often teaming up to increase the fun.

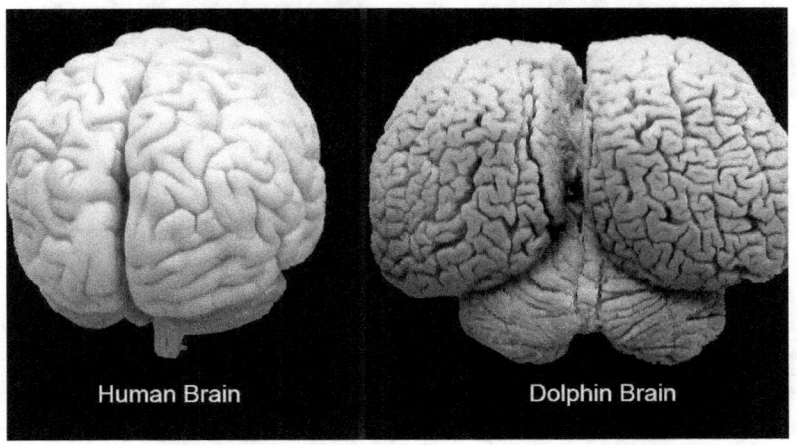

Human Brain   Dolphin Brain

Probably the most important reason why they are considered intelligent; dolphins click and whistle. They can send pulses of loud broadband packets of sounds, which they use to communicate, ward off predators, discipline the young, etc.

119

Whales, a part of their family, "sing". All these have baffled scientists and have not yet deciphered what each and every "packet" of sound means.

Does that mean that in time they could build a rocket and send a dolphin to the moon? Judging by their physical appearance, and the fact they live underwater, we can safely assume that this is very unlikely. So not only does a species need intelligence, it also needs to live on land, have the dexterity to manipulate objects, and the desire to build a civilisation with specialised jobs, write everything down to pass on to future generations so each new born doesn't have to start all over. Intelligence therefore requires many other aspects to travel into space or at least learn how to communicate across the universe.

Dolphin show at the Mirage, Las Vegas. Photo by the author.

Other primates that look rather like us do indeed occupy our planet at the same time as ourselves. If we drew a timeline of the history of our planet with each millimetre representing a million years, then the history of the Earth would be a chart 4.6 metres long. All the primates in the history of the Earth are all on the last 5mm of that chart. Draw it in the real world to see the significance. Is this a coincidence?

We share around 98% of the DNA that all other primates have around the globe, we even have the same blood types. Gorillas are of type B & O while Chimpanzees are of type A & O; another coincidence? Off course not, Charles Darwin came to the correct conclusion that we had to be related even though DNA had not been discovered during his lifetime.

"Was probable that Africa was inhabited by extinct apes closely allied to the gorilla and the chimpanzee; and as these two species are man's nearest allies, it is somewhat more probable that our early progenitors lived on the African continent." Charles Darwin.

So not only have we one species developing technology, the Earth came very close to producing several other species capable of building rockets at the same time. Only a 30% lower I.Q within the others prevented this from occurring.

Further evidence of the potential of other primates was clearly demonstrated by some very special experiments. Sign language has been taught to some Gorillas and Chimpanzees. The most famous of which is probably Koko.

We often ask ourselves *"Is there intelligent life out there?"*

Perhaps we should rephrase it to -

"Are there intelligent technological civilisations out there?"

This would be a far more precise question.

Chapter 25 Technological Growth

Regarding the Human Race, technology began with the very first items to make life more bearable and survivable; hammers made of flat stone, knives from flint, rope from vines and clothing from animal skins. This level of technology remained for centuries. Homes moved on from caves to mud huts where they could be built almost anywhere and expanded into villages. Fences kept out wild animals and improve overall security. Each member of the community took on a vital role and teams / jobs were created to produce a more efficient way of life. Writing and language helped to pass on stories, inventions and discoveries from one generation to the next so they did not have to start all over.

Once spare time was on a humans' side, we began to ask questions about nature and how it worked. Exploration also became a natural progression and mapping / navigational techniques were developed to accelerate the process.

An Iron Age replica of a home at Flag Fen;
Near Peterborough, England. Photo by the author.

The process of this thinking was intricately tied with mythology, early religious beliefs and became bogged down with the hierarchy of harsh leaders, kings and priests. It is not inevitable that technology will develop within some cultures.

A classic example is that of the Australian Aborigines. As far as the research suggests, they have occupied the continent for around 50,000 years and made no progress beyond huts, spears, log fires etc.

Progress toward modern technology was slow even outside Australia until a new way of thinking developed. This arguably began with the Polish revolutionary **Nicolas Copernicus** (1473-1543). He deliberately separated his thoughts from any faith or myth and observed the sky as it was rather than as it should be. It was always felt that the Earth was of total importance to us therefore it must be at the centre of the Universe.

He observed the paths of the planets and realised that they would make perfect sense if they orbited the Sun instead of the earth. This did not quite match the observed sky but was a lot simpler than trying to explain the paths using circles within circles. This was the earlier Ptolemaic system that had been adopted by most faiths for centuries that put Earth at the centre. On his deathbed in 1543, Copernicus asked for his theory to be published using his own money. He knew that any earlier announcement of this idea would have resulted in being burned at the stake for blasphemy.

Galileo Galilei was born in 1564 at Pizza, Italy. As a mathematician, he was fascinated by the Copernican model of the solar system and later set out to prove the correct answer. The invention of the telescope in 1608 was the key to this success. He built his own telescopes, the first had a measly magnification of 3x, the next was 30x. He observed everything he could in the sky and produced sketches, observations and measurements.

On 7 January 1610, Galileo observed with his telescope what he described at the time as "three fixed stars, totally invisible by their smallness", all close to Jupiter, and lying on a straight line through it. Observations on subsequent nights showed that the positions of these "stars" relative to Jupiter were changing in a way that would have been inexplicable if they had really been fixed stars. On 10 January, Galileo noted that one of them had disappeared, an observation which he attributed to its being hidden behind Jupiter. Within a few days, he concluded that they were orbiting Jupiter: he had discovered three of the four largest moons. He discovered the fourth on 13 January, just visible in his homemade telescope. This one was probably Io being the faintest of the four.

His discovery of moons orbiting Jupiter was the single most disturbing event in the eyes of the Catholic Church. There were objects orbiting another body other than the Earth. The Earth was not the only centre of importance; another body is at a centre of its own system. The Catholic Church put Galileo under house arrest and ordered him to keep his observations to himself. They even refused to look through his telescope, as they believed it to be haunted by demons.

Then he discovered blemishes on the sun; sunspots. These seem to come and go and demonstrated that the Sun rotated. It was not a perfect unblemished sphere that provided life on

Earth through a spiritual being, but was a physical entity operating and producing energy by unknown physical laws instead.

Isaac Newton (1642-1727) accelerated the new way of thinking about our universe. He proved beyond doubt that universal laws existed to control everything from the flight of a canon ball to a planet. Mathematics could be used to predict such motions. His discovery that white light was made up of varying frequencies released the new science of Spectroscopy and allowed us to discover what stars were made of without going to one. All this led to an unstoppable revolution in technology and knowledge.

Isaac Newton's home in Woolsthorpe, Lincolnshire, UK.

Speed: The sheer speed of technological growth in the 19th and 20th centuries as a result of viewing the world in a realistic manner has been phenomenal so far although some resist the movement. The chief engineer of the British Post Office after hearing of the invention of the telephone once said, "The American's have need of the telephone, but we do not. We have plenty of messenger boys."

Others are often caught out in the opposite direction. An American Mayor was once quoted saying, *"We shall have a telephone in every village. I believe it is a realistic desirable goal by the year 2000."* Little did he realise we would even have satellites carrying millions of human conversations at once across the planet in real time long before 2000.

Many scientists today are still being challenged about the morals of new technologies. Some people still strive to slow down new knowledge, inventions and exploration. Wernher Von Braun in 1969, chief rocket designer at NASA just before the launch of Apollo 11 to the Moon was approached by a preacher, *'If we were meant to walk on the Moon then God would have given us rockets on our feet.'* His reply was *'If he didn't want us to land on the moon, he shouldn't have given us the brains to do it.'*

The future is our choice and is not destined to succeed or fail.

20th Century: As an example of the rate of progress, we can choose a particular area and list a few key points. Many people witnessed a series of events in one single lifetime. We will take flight as a perfect example.

Leonardo Da Vinci began the process as far back as 1485 when he correctly designed a wing and even helicopter rotors. The first truly powered flight of a heavier than air machine was in 1903 by the Wright Brothers. The aircraft was homemade, built in their bicycle workshop and flew in Kitty Hawk, North Carolina.

The Flyer – first Wright Brothers aircraft at the Smithsonian Museum, Washington DC. Photo by the Author 1990.

By 1914, aircraft were used in the First World War for surveillance and then for dropping bombs by reaching over the side of the plane and literally letting go of a bomb by hand.

Charles Lindbergh flew non-stop from New York to Paris in 1927 in his famous aircraft Spirit of St Louis. By 1947, we witnessed the first aircraft to break the Sound Barrier. Then in 1961, Yuri Gagarin orbited the Earth in a spacecraft at 17,000mph and survived. 1969, the first Moon landing occurred with Neil Armstrong & Buzz Aldrin exploring the dusty surface, human footprints on another world.

The Apollo 11 Spacecraft in Washington DC.
Photo by the author 2016.

If we take into account-unmanned flights, 1987 saw the first spacecraft fly past the orbit of Pluto - Pioneer 10. This was the first of five probes so far to leave the old definition of the solar system.

From the first powered flight in 1903 of just 12 feet or so altitude, to achieving a distance of 4 billion miles in just 84 years is truly staggering. At the same rate of progress, we should have the first space probe heading for the stars by around 2040 and reach one within a single human lifespan. It is possible that before its arrival, the first human crews will be heading that way too and even perhaps overtakes the first probe.

Types of Civilisations: There's something called The Kardaashev Scale, which helps us group intelligent civilizations into three broad categories by the amount of energy they use rather than rate of progress:

Type I has the ability to use all of the energy on their planet. We are not quite a Type I Civilisation, but we are close (Carl Sagan created a formula for this scale which puts us at a Type 0.7 Civilization).

Type II can harness all of the energy of their host star. Our feeble Type I brains can hardly imagine how someone would do this, but we have tried our best, imagining things like a Dyson Sphere. These are rings around a star of solar panels. They absorb energy from the star and transfer it to the home planet and other colonies. A single ring can be increased to many rings at other angles as power demand increases. The project could indeed engulf the entire star as a sphere of solar panels absorbing the entire output.

Type III accessing power comparable to that of the entire Milky Way. If this level of technology sounds hard to believe, remember the universe is old enough to produce them. If a civilisation similar to ours were able to survive to Type III level, the thought is that they would probably have mastered inter-stellar travel, possibly even colonizing the entire galaxy.

128

One hypothesis as to how galactic colonization could happen is by creating spacecraft that can travel to other planets, spend perhaps 500 years or so self-replicating using the raw materials on their new planet, and then send two replicas off to do the same thing. Even without travelling anywhere near light speed, this species would colonize the whole galaxy in 3.75 million years - that is fast compared to the age of the galaxy of at least 12 billion years.

If we allow 5 billion years for the first civilisation to appear that could achieve this, then it means the galaxy is old enough for this scenario to have occurred 1,866 times over if just one species existed at a time. Our estimate of civilisations from the chapter on the Frank Equation is correct at 40,500 then the number is then 75,573,000 species sending out millions of probes for the past 7 billion years. We haven't found a trace of one; we do have a problem don't we?

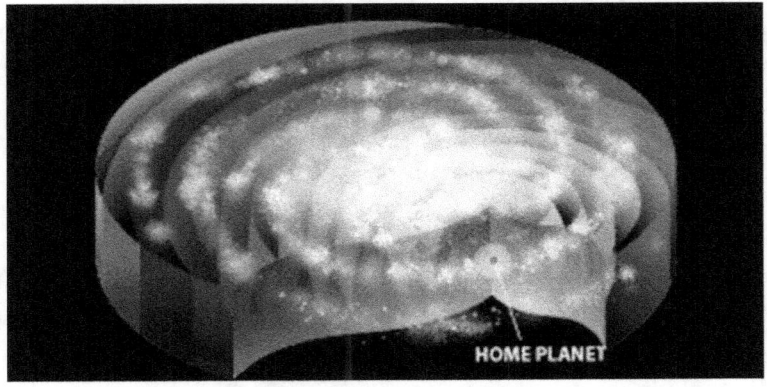

HOME PLANET

Even if 1% of intelligent life survives long enough to become a galaxy-colonizing Type III Civilization, then there should be at least 1,000 Type III Civilizations in the Milky Way alone— and given the power of such a civilization, their presence would be very obvious by now. Yet, we see nothing, hear nothing, and no one visits us as far as we know.

Note we do seem to keep coming back to this feeling of being alone in this vast galaxy of some 200 billion stars.

Chapter 26   A Choice of Futures

If the human race is to eventually break away and set up permanent homes in space, then a stable launch pad from Earth is required. Our civilisation itself will require surplus funds, materials, and a long term political will to design and pay for an intricate system for a chosen few to venture out into space and never return. Only this action will guarantee long-term survival for countless millions of years.

Risk: All the while, we are stuck here on this little blue sphere; we are prone to suffer the fate of the dinosaurs in many ways. They all died out due to an asteroid impact. We are currently living with the same risk. Even though we have our intelligence and a reasonably advanced technology base, a similar strike could occur with just a few days' notice. There would be nothing we could do to prevent the impact. Perhaps a small handful of citizens could survive underground for a few months then try to scrape a living out of the soil after the weather had settled down somewhat and the skies cleared again. Such events are rare but statically we are overdue for a similar sized impact by around 5 million years.

Many experts agree that the biggest risk we face is from a virus. Movies such as Stephen King's 'The Stand', 'Outbreak' from 1995, 'I am Legend' in 2007, 'Twelve Monkeys' with Brad Pitt and not forgetting 'The Andromeda Strain' written in 1969 by Michael Crichton, all demonstrate the potential disaster we face at any one moment. With increasing interaction with animals, biological experiments, and climate change that is allowing various creatures to live in places never populated before, the risk we face is actually increasing.

Our intelligence can indeed solve our potential extinction but can also cause it. Even into the 21st Century, after many nuclear arms reductions, we still hold enough active nuclear weapons to destroy the surface of the planet many times over. Some people believe a limited nuclear exchange can be contained politically and is survivable in such a region. Chaos and destruction would rein for some time but would be survivable.

Even though there is enough firepower to destroy, and lay radioactive waste to every square mile of land, this is unlikely to happen unless every weapon were released on every single continent. Perhaps a nuclear winter lasting many years would occur as predicted by Dr Carl Sagan and even the survivors of such a war will then face the hardship of starvation and radiation sickness. The movie 'The Day After' released in 1983 is a perfect example of this scenario and then perhaps another called *'Damnation Alley'* with George Peppard although a little farfetched with Earth tilting over, then back again.

Such possible depressing scenes are in fact amongst the realistic possible futures we face. Such risks can be reduced to zero or virtually zero if we choose. The only single answer is to send out self-sustaining colonies into deep space. They would never need materials or finances from Earth once they leave. If they were to set up on Mars for instance, then any disease roaming around here can be halted from reaching Mars. A natural 6-month period is required to reach the planet and anyone leaving Earth in haste that is a carrier of such a deadly virus, there is plenty of time to screen them and isolate and even refuse landing facilities etc. An entire protocol needs to be defined for such a possible scene.

A nuclear war on Earth will not involve a self-supporting space colony. A drama series in the 1970s has called 'The Martian Chronicles' by Ray Bradbury depicted just that. A major conflict broke out and asked the Mars residents to return to Earth and carry out their duties. They all refused and watched the Earth turn into a radioactive planet through their telescopes. Again, a depressing message but serves as a warning to give us the chance to learn that such scenes are indeed possible and have already come close to reality several times.

In 1908, a small comet entered the earth's atmosphere and destroyed a large part of the Siberian forest. Several million trees were blown down radially from the air blast. The dust altered the weather across Europe for weeks and cooled the atmosphere. Spectacular sunsets were observed in London. So much light was being reflected by particles in the upper atmosphere from the comet, that the night just did not happen until much later each evening.

Several asteroids of a similar size to the dinosaur killer have passed dangerously close to us and have even altered their regular solar orbits due to our gravity. An actual impact of such a magnitude is incredibly rare, but a glance at the Moon through a telescope demonstrates clearly that it occurs more often than one may realise if a geological time-scale is used.

Do not forget the Earth has six times the strength of gravity and is four times the moon's diameter. The Earth has been hit more often than the moon. Natural wind & rain erosion, the oceans and plate tectonics have destroyed most of impacts from history, so the lack of craters gives us a false sense of security.

Meteor Crater, Arizona. A 50-metre-wide chunk of Iron gouged out this crater some 50,000 years ago. The original floor is deeper; dust blows in annually and is slowly filling the crater. Photo by the author in 2006.

Viruses such as Ebola, Cholera, SARS or even strains of the flu have the potential of killing millions of people within a few weeks. Our modern mass air travel would be the almost unstoppable mechanism that puts us all at risk. Very strict protocols are set in place whenever such outbreaks occur such as infrared face images at airports or instant temperature and perspiration checks etc.

The biggest pandemic in history is the 1918 flu outbreak caused by soldiers returning home after hardship in the trenches. In total, 75 million people died, several times more than that in the First World War itself and yet the event is largely forgotten. HIV / Aids have so far claimed 30 million lives. The Coronavirus has claimed 2.1 million lives as of March 2021.

The year 1962 will always be remembered as the time when we came closer than ever to destroying ourselves. On the morning of 16 October, National Security Advisor McGeorge Bundy informed President John F. Kennedy that U.S. surveillance aircraft had discovered the presence of Soviet missiles in Cuba, just 90 miles from American soil. It was the start of the Cuban Missile Crisis, which brought the world to the brink of nuclear war.

The U-2 aerial photographs were analysed inside a secret office. The critical photographs snapped by planes over Cuba were shipped for analysis to a top-secret CIA facility in a most unlikely location: a building above the Steuart Ford car dealership in a rundown section of Washington, D.C. While car sales representatives were wheeling and dealing downstairs, upstairs CIA analysts in the National Photographic Interpretation Center were working around the clock to scour hundreds of grainy photographs for evidence of a Soviet ballistic missile site under construction.

Beginning in the summer of 1962, the Soviets employed an elaborate operation code-named Anadyr, to ship thousands of combat troops to Cuba. Soldiers donned chequered shirts to pose as civilian agricultural advisers. Many more were issued Arctic equipment to throw off the scent, sent aboard 85 various ships and then told to remain below for the long voyage. When the CIA estimated on 20 October that between 6000-8000 Soviet troops were stationed in Cuba, the true number was more than 40,000 - revealed after the collapse of the Soviet Union.

In a dramatic address on 22 October 1962, Kennedy informed the nation of the naval blockade around Cuba. An alternative speech with a much different message had been drafted days before, however, in the event the president opted for a military strike. *'This morning, I reluctantly ordered the armed forces to attack and destroy the nuclear build up in Cuba'* began the address that JFK never delivered.

John F. Kennedy delivering his ultimatum to the Soviet Union

On 27 October, a Soviet-supplied surface-to-air missile downed an American U-2 plane, killing its pilot; Major Rudolf Anderson Jr. President Kennedy posthumously awarded him the Distinguished Service Medal.

U.S. Secretary of State Dean Rusk said of the Cuban Missile Crisis, *'We're eyeball to eyeball, and I think the other fellow just blinked.'* That assessment is too one-sided. While on 28 October, Soviet leader Nikita Khrushchev ordered the removal of Soviet nuclear missiles from Cuba and the Americans also secretly pledged to withdraw intermediate nuclear missiles from Turkey.

134

As clearly demonstrated, we have a choice to define our own future. We could bury our heads in the sand and ignore the asteroid threat, not bother researching cures for new diseases, or make hasty and poor decisions in war. Better still, we can learn from the past and invest time, resources, money and wisdom at preventing such disasters from occurring. If we choose the latter soon enough, we will inevitably spread ourselves ever further across space and make our species immortal.

Climate Change: Details of Climate Change has been covered in many forms, there is no need for this book to include a repeat. As from 2020, we already know that the 2010's has been by far the warmest decade on record. The UK has announced a massive goal of becoming Carbon Neutral by 2050. If we take a global view, and assume the target has already been reached today, skip out 30 years, this means the total emissions from our planet has dropped by 1%. This will be counter-acted and reversed by China alone within 2 years.

It is one point to have an ambitious target for a relatively small nation and set an example, but in reality, unless other larger nations follow immediately, the exercise becomes a meaningless gesture.

In the meantime, our population grows, demands on material resources and energy increases, climate change accelerates. Demand on the natural the world that remains, damages it beyond repair, and sets in motion a mass extinction beyond most similar natural events from the past.

With our science, we can combat most diseases and other forms of population reduction. We need to continue to grow greater volumes of food by destroying more of our own carbon dioxide absorbing forests and expand this never end ending destructive cycle.

Nature can only accept a certain level of damage before a complete collapse. Our population then would crash within a single generation, hence our technological base too and any chance of breaking away from this planet.

Inevitable Conclusion? If other civilisations run the same risks as we, perhaps it is inevitable that technological species such as ours have a very short span. This could be a possible answer as to why the sky is not crowded with alien signals and artefacts from space faring species.

However, it would only take one smart and wise group of beings that truly learns from their own past and strive beyond the point we are currently at. They would move away from their place of origin just as we potentially could. That one species could hop from one planet to the next, one star system to the next and spread out across the galaxy. They could even build mini-planets in the shape of a cylinder, rotate it for artificial gravity; massive artificial homes for thousands as mentioned earlier.

There are no laws of physics, chemistry, or biology that could even prevent such a scenario from unfolding. It may take a million years or so for the Milky Way to become colonised by a single species, but it is billions of years old; plenty of time for such a race to do it. So once again –

Where are they all?

Chapter 27   Final Numbers for the Drake Equation with current Knowledge

When the Drake Equation was first published, many of the numbers that need to be inserted were completely unknown. Wild estimates of a maximum and minimum range were inserted so that the resulting figures of how many civilisations out there were extremely vague.

Since around 1990, the technology has been developed to look much more closely at most aspects and even update the formula. We can re-write it accurately with a minimum and maximum range...

$N = Ns \times Fp \times Ne \times Fl \times Fi \times Fc \times Fl$

Ns = Number of stars in the Milky Way 200 billion

Fp = fraction of stars with planets, 50%, 100 billion

Ne = number of planets plus moons that could sustain life, 3.

Fl = fraction of those bodies where life does evolve, 0-3

Fi = fraction of those where intelligent life evolves and create technology, 5%?

Fc = fraction of those that develop long distance communications across space, 90%

Fl = fraction of civilisations surviving lifespan from first transmissions, 100 yrs (99%?) to millions of years (10%?).

From these numbers, it can be calculated that the minimum number of civilisations currently in the Milky Way apart from ourselves (N) is a minimum of zero to a maximum of **40,500.** This does not include the strong possibility of just one of them colonising the rest of the galaxy. If this has happened then the sky would be screaming with artificial signals, so why isn't it?

MUSINGS ON THE PROBABILITY OF ALIEN LIFE EXISTING SOMEWHERE IN THE UNIVERSE:

THERE ARE TRILLIONS OF GALAXIES

.. EACH WITH BILLIONS OF STARS

IN THE OBSERVABLE UNIVERSE

SURELY SOME OF THEM HARBOR LIFE

OUR GALAXY ALONE HAS 400 BILLION!

IN FACT: THERE MAY BE 2 PLANETS PER STAR

YET STARS ARE OUTBUMBERED BY PLANETS...

OF THEM, 20 BILLION ARE "EARTH-LIKE"

AND LESS THAN 5,000 ARE KNOWN!

HOW CAN ANYONE EVER SAY WE ARE ALONE?

Kepler Search Space
3,000 light years

Sun

Image Credit: NASA (Top Left), Steve Mandel/John Gleason(Milky Way), Shutterstock (Center Right), Jon Lomberg (Bottom Left), NASA/JPL (Bottom Right)

The latest and most accurate number of technology-based civilisations in the Milky Way at any one time should be around 40,500.

Chapter 28 The Fermi Paradox

Enrico Fermi was born in Italy, 1901. At school, he excelled in mathematics and took an interest in Physics and Astronomy. Enrico loved to examine big intriguing thoughts. Frank Drake's famous equation troubled him somewhat. With the estimated figures placed within the formula, thousands of civilisations should exist within the Milky Way alone at any one time.

We have already demonstrated that once self-sustaining colonies are set up around a star system, then no disaster could ever wipe out the species. In time, any one species can spread out across the entire galaxy; no physical laws can prevent it. Therefore, given the age of the galaxy, there should be aliens all over the place, everywhere we listen for radio signals, there they should be. After examining over one hundred million stars on thousands of channels for radio transmissions – Nothing! Not even a whisper. Why? This is the scenario proposed by Enrico Fermi.

There just may be a sensible answer as to why evidence of their presence is missing from the radio sky. Here a few listed possible solutions...

Signals have not had time to reach us: Some would have had time, and others would not. At light speed if an alien transmitted anything from 1000 light years away but only pushed the transmit button in 1816, then we then we have to wait another 800 years to receive it. Then again, if another transmitted a more powerful signal from 10,000 light years away 10,000 years ago then died out, we would still receive it today. The time delay makes no difference into finding some signals; all we need is one!

Data not understood: Perhaps they are transmitting across the galaxy but in a way that our detectors are not programmed for. This is the job for the technicians that operate SETI projects around the globe.

Universal Language: If many civilisations do exist, perhaps they have a universal agreement to keep the 'chatter' restricted in directional transmissions so become unlikely to be picked up from Earth or any other planet looking for the same as we. The reason could be to keep youngsters like ourselves asking questions about the cosmos until a more mature / advanced stage is reached.

Others in a similar position as ourselves would not have reached that level yet and would still be broadcasting to the galaxy as we are now. These examples are the ones we should be hearing now, but as of 2020, the silence remain in place.

Inevitable Extinction? This is probably the most depressing of all scenarios. As we still only have the one example to analyse, we have no idea that this could be our fate too. In the meantime, we are still broadcasting; those signals can never be halted. If we continue to broadcast radio until say 2050, then over 100 years' worth of news expands in a never halting bubble across the galaxy. Any aliens 1000 light years away will not hear of us yet. However, by our date of 2939, they will hear of the outbreak of World War 2. They will continue to hear us for more than a century until the last transmission; then silence.

If this is the case then the chances of hearing from others are remote, we need to be listening at the right place in the sky at the right time when their 'bubble' broadcast reaches us. In addition, such cases would mean that the transmissions might not be very powerful as in our case too.

Faint Signals: If civilisations are just rare for some reason and never depart in a big way from their home planet, then there may be no reason to transmit powerfully across space. Such a signal trace would be incredibly faint as ours would be. Perhaps we are receiving them but they become obscured by all the natural background noise from the cosmos.

Perhaps such systems as this one in New Mexico is not powerful enough to receive faint signals from extra-terrestrials yet. Photo by the author 2013.

Conspiracies: Perhaps signals / complete flying saucers have been found but governments are covering it up. This could work to a degree, but radio telescopes are mostly in the hands of astronomers, not working for governments. They do not have a monopoly of radio signals from space and cannot stop such information reaching the public domain. The *'Wow'* signal was never kept secret as a powerful example (see *Wow Signal Chapter*).

Life, but that is it! We are naturally obsessed with finding creatures like ourselves to communicate with. Perhaps the development of the Human species was a unique turn of events; a very rare one-off occurrence by accident or design by a spiritual being.

Some 50 billion species have arisen in one form or another on Earth throughout its entire biological history of over 3 billion years. Only one has ever produced technology. *So on Earth intelligent life is a very rare event in itself and not inevitable.*

This does not stop the possibility of life in general existing elsewhere. The numbers for the Milky Way alone are in the billions of possible inhabited life bearing worlds.

There are almost certainly other reasons to explore. We need to keep listening, tuned into the sky with our powerful receivers, and forever improve our detection methods. It is bad science to assume we are alone or part of a crowd of civilisations, but good science to try to reveal the truth.

Guessing or assuming is what kept us in the dark ages. We lost much of our population through treating the Plague via witchcraft for instance instead of inventing the microscope and study the bacteria.

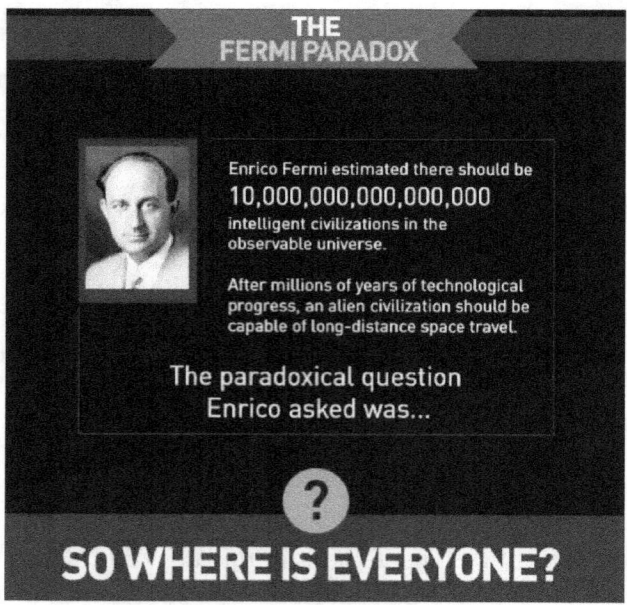

To experiment with your own numbers, a link can be found on the website <u>www.outerspacebooks.com</u> *on the Frank Drake Equation page.*

Chapter 29   The Arecibo Message

In 1974, the most powerful broadcast ever deliberately beamed into space was from Puerto Rico. The broadcast formed part of the ceremonies held to mark a major upgrade to the Arecibo Radio Telescope. The transmission consisted of a pictorial message, aimed at the globular star cluster M13. This star cluster is roughly 21,000 light-years from us, near the edge of the Milky Way galaxy, and contains approximately a third of a million stars.

The broadcast was particularly powerful as it used Arecibo megawatt transmitter attached to its 1000-foot-wide antenna. It concentrated the transmitter energy by beaming it into a very small patch of sky. The emission was equivalent to a 20 trillion-watt single direction broadcast. It would be detectable by a similar SETI experiment by another civilisation just about anywhere in the galaxy, assuming a receiver is similar in size to Arecibo.

The message consists of 1679 bits, arranged into 73 lines of 23 characters per line (these are both prime numbers, and may help the aliens decode the message). The same theory was demonstrated in the movie 'Contact.'

The "ones" and "zeroes" were transmitted by frequency shifting at the rate of 10 bits per second. The total broadcast was less than three minutes in duration. A graphic showing the message is reproduced here. It consists, among other things, the Arecibo telescope, our solar system, DNA, a stick figure of a human, and some of the bio-chemicals of earthly life. Although it is unlikely that this short inquiry will ever prompt a reply, the experiment was useful in getting us to think a bit about the difficulties of communicating across space, time, and with another language.

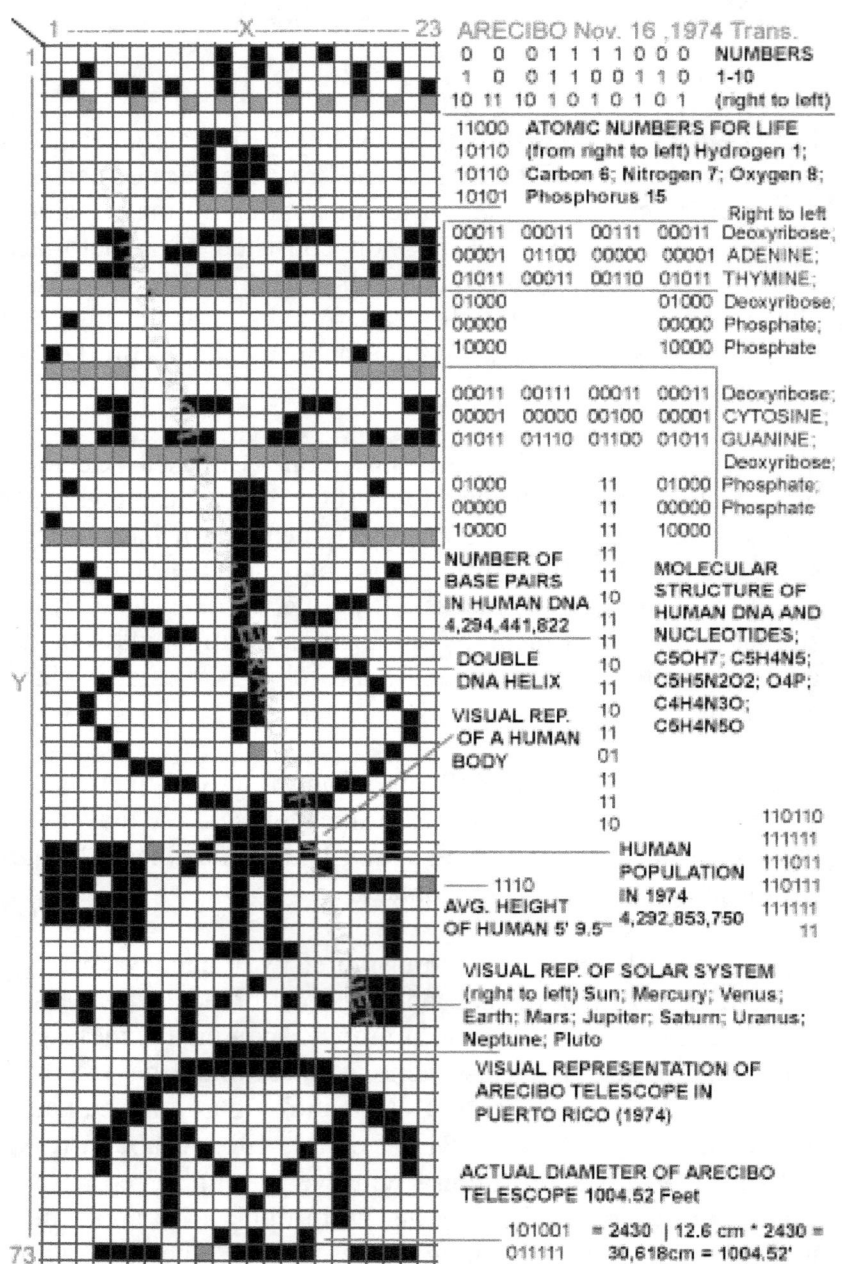

ARECIBO Nov. 16 ,1974 Trans.

0	0	0	1	1	1	1	0	0	0		**NUMBERS**
1	0	0	1	1	0	0	1	1	0		1-10
10	11	10	1	0	1	0	1	0	1		(right to left)

11000   ATOMIC NUMBERS FOR LIFE
10110   (from right to left) Hydrogen 1;
10110   Carbon 6; Nitrogen 7; Oxygen 8;
10101   Phosphorus 15

Right to left

00011	00011	00111	00011	Deoxyribose;
00001	01100	00000	00001	ADENINE;
01011	00011	00110	01011	THYMINE;
01000			01000	Deoxyribose;
00000			00000	Phosphate;
10000			10000	Phosphate

00011	00111	00011	00011	Deoxyribose;
00001	00000	00100	00001	CYTOSINE;
01011	01110	01100	01011	GUANINE;
				Deoxyribose;
01000	11		01000	Phosphate;
00000	11		00000	Phosphate
10000	11		10000	

NUMBER OF
BASE PAIRS
IN HUMAN DNA
4,294,441,822

DOUBLE
DNA HELIX

VISUAL REP.
OF A HUMAN
BODY

MOLECULAR
STRUCTURE OF
HUMAN DNA AND
NUCLEOTIDES;

C_5OH_7; $C_5H_4N_5$;
$C_5H_5N_2O_2$; O_4P;
$C_4H_4N_3O$;
$C_5H_4N_5O$

11
11
11
10
11
11
10
11
10
11
01
11
11
10

110110
111111
111011
110111
111111
11

HUMAN
POPULATION
IN 1974
4,292,853,750

1110
AVG. HEIGHT
OF HUMAN 5' 9.5"

VISUAL REP. OF SOLAR SYSTEM
(right to left) Sun; Mercury; Venus;
Earth; Mars; Jupiter; Saturn; Uranus;
Neptune; Pluto

VISUAL REPRESENTATION OF
ARECIBO TELESCOPE IN
PUERTO RICO (1974)

ACTUAL DIAMETER OF ARECIBO
TELESCOPE 1004.52 Feet

101001   = 2430 | 12.6 cm * 2430 =
011111   30,618cm = 1004.52'

144

Chapter 30   Our Probes into the abyss

The first spacecraft to be sent away from the Earth was the Russian Luna 1 launched on 4 January 1959. It reached within 6,000km of the moon. It flew straight past and entered an orbit around the sun. The first US probe to perform the same was NASA's Pioneer 4. After launch in March 1959, it was intended to fly past the Moon up close and send data on the radiation in its vicinity. The trajectory missed the Moon by 60,000km; it did send back some useful data but ceased transmissions after 82 hrs.

Such probes became more reliable and ambitious. In 1972 and 1973, NASA launched the first probes toward the outer solar system; Pioneer 10 & 11. These were the first to be accelerated so much by Jupiter's and then Saturn's gravity, that the final velocity exceeded that to which the Sun can hold on to them. The sun's gravity would simply not be strong enough at that distance to bring them back. They were the first destined for the stars. Their own momentum will take our instruments out into the galaxy regardless of what happens here on Earth in the near or distant future.

The next probes to head for the stars were Voyager 1 & 2. Both launched in 1977, they headed for Jupiter then onto Saturn. Voyager 1 approached Saturn from a low angle and then flew over Saturn's North Pole to obtain unique perspective. However, the final trajectory means is will continue to head upward away from the other planets. Voyager 2 remained on the same plane after Saturn and was able to swing past to Uranus in 1986 and then Neptune in 1989. After flying past four planets, all pulling on the craft accelerating it faster and faster, it became the fastest spacecraft ever at 38,000mph.

Voyager 2 is streaking toward an encounter with a star called AC +793 888, which currently lies 17.6 light-years away. Voyager's on its way to a close approach with it in 40,000 years taking into account the star is moving roughly toward us to help get there faster. It will swing by it, and it will then continue to orbit around the centre of our Milky Way galaxy.

New Horizons is an interplanetary space probe that was launched as a part of NASA's New Frontiers Program. The spacecraft was launched in 2006 from Florida with the primary mission to fly past and study the Pluto system, and a hopefully a secondary mission to fly by and study at least one more Kuiper Belt object (KBO).

It was launched to speed of about 36,373 mph. After a brief encounter with asteroid 132524APL, the probe proceeded on to Jupiter, making its closest approach on 28 February 2007, at a distance of 1.4 million miles. Jupiter's gravity freely provided extra speed enabling it to reach Pluto on 14 July 2015. It flew 7,800 miles above the surface and became the first spacecraft to explore the dwarf planet. On 25 October 2016, the last of the recorded data from the Pluto flyby was received from New Horizons. Having completed the first phase of its mission, another Kuiper Belt object was chosen to be studied as it approaches in 2019. From then on, it will become the fifth probe to head out for the stars. These examples simply prove beyond doubt that reaching the stars is possible from a technical viewpoint, albeit taking thousands of years. All that is required to make such journeys more practical is to improve the velocity to reduce journey times on a more human timescale. Such probes are only travelling at 0.000058 of the speed of light.

New Horizons Accelerating toward Jupiter for its main gravity assist in 2006. This extra velocity gave it enough energy to head for Pluto and then become the fifth craft to escape from the solar system completely.

As the mission planners knew the ultimate destiny of these amazing machines, various forms of messages were built onto the craft to show any aliens that just may come across one where they came from and who built them.

Pioneer 10 & 11 had a gold plaque clasped to the outside of the craft that included a map of where the solar system is compared to several neutron stars around the Milky Way. It should not be too difficult for any extra-terrestrials to work out where we are or perhaps were. It also included a rather rude picture of a man and woman so I have refrained from showing it for rather obvious reasons. Perhaps any aliens finding one of the probes will get the impression we have not invented clothes. It was designed by were designed by Dr Carl Sagan and Dr Frank Drake and drawn by Dr Sagan's wife Linda. The Pioneer plaques used the hydrogen atom as a basis for describing the location of our solar system in the Milky Way and the relative locations of the planets in our solar system.

Voyager 1 & 2 spacecraft carry more subtle messages to the cosmos. Ann Dryan, a NASA scientist was the creative director of the Voyager message. The most famous part was a record that can be played to hear human voices and music. The musical choices included works by Bach, Mozart, Louis Armstrong, Chuck Berry, and a Pow Wow performance by the Navajo Native Americans.

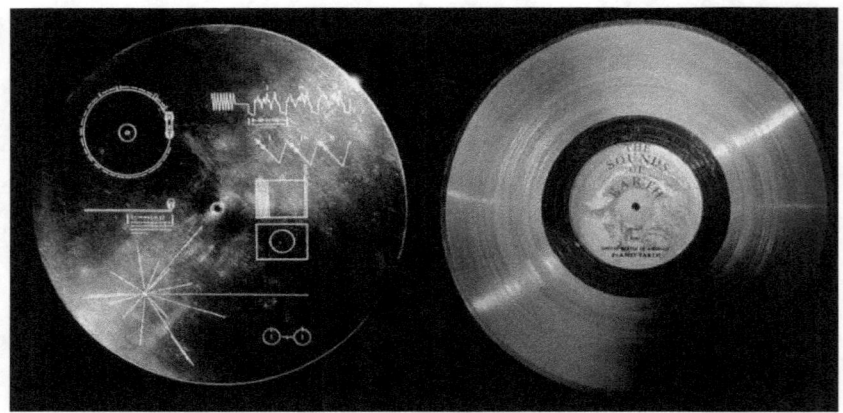

The Voyager Record; left above is exposed on the outer spacecraft; the other was protected from micro-meteorites hitting the craft and destroying the data. NASA

Many photographic slides were recorded digitally on the record. The images included factories, a baby, cities, the Earth from space and a picture of a naked man and woman. Well it was to be included but NASA pulled it out at the last minute. After launching Pioneer 10 & 11 with naked human pictures, they did not want to give the impression that we were obsessed with nudity.

A detailed book on the making of this project is 'Murmurs of Earth' by Carl Sagan, Ann Dryan, Linda Sagan and Frank Drake.

Chapter 31   The 'Wow!' Signal

On 15 August 1977, a radio signal was received at the Big Ear radio observatory in Ohio. It was no ordinary signal. The volunteer astronomer who spotted the pattern on the paper logs circled the data and wrote "Wow!" in the margin. The radio telescope was observing space as part of the Search for Extra-Terrestrial Intelligence (SETI) program, and it was the most promising signal the receiver had recorded in its fourteen years of full operation. It was powerful enough to push the monitors off the limit.

The signal came from constellation Sagittarius, and lasted seventy-two seconds at about 1420.456 MHz before it faded away. The volunteer who found and circled the data in the paper printout was Jerry Ehman, who was amazed at the signal's intensity and what a narrow range of frequencies it appeared in. Seventy-two seconds also happened to be the exact length of time it would take the Earth to rotate the Big Ear through a signal from space. He did some analysis of the data, and by all indications this powerful, narrowband radio signal was from outside of our solar system. However, was it sent by an advanced civilization?

Curiously, only one of the scope's two detectors picked up the signal. When the second detector covered the same patch of sky three minutes later, it heard nothing. This indicated the unlikely possibility either that the first beam had detected something that was not there, or that the source of the signal had been shut off or redirected in the intervening time. The observatory researchers trained their massive scope on that part of the sky for a full month, watching closely for a repeat of the mysterious signal. Nothing interesting was observed during those thirty days, yet scientists were at a loss for an explanation of the original event. Planning to return to that patch of sky periodically, the Big Ear continued its broader purpose.

Above; the Big Ear receiver horn.
Photos in this chapter all by the author 1992

Several times over the next twenty years, respected SETI researcher Robert Gray and his colleague Kevin B. Marvel arranged for further scans of that region of space. They managed to obtain some time on the META array at the Oak Ridge Observatory in Massachusetts, and the extremely sensitive Very Large Array (VLA) in New Mexico, made up of twenty-seven 25-meter-wide radio dishes. They detected some extremely faint sources of radio emissions in the infamous patch of sky, but nothing like that of the "Wow!" signal.

Their findings did essentially disprove the only working theory as to the cause of the original event: *'interstellar scintillation.'* It was thought that perhaps some weaker radio signal from space had been temporarily focused on the Big Ear in a way similar to stars twinkling… but the VLA is sensitive enough that it would have detected such a source; it did not.

The Very Large Array (VLA) in New Mexico. A powerful telescope that was used to re-allocate the 'Wow' signal without success.

We chatted about the 'Wow!' signal to a technician, Larry Brothers at the VLA itself.

The Big Ear maintained its periodic scan of that part of space for almost forty years, and never again came across such a compelling signal. 'Wow' remains the clearest signal ever received from an unknown source in space, as well as the most fascinating and unexplainable.

The signal's original discoverer Jerry Ehman does not care to speculate on its source and remains scientifically sceptical. *'Even if it were intelligent beings sending a signal, they'd do it far more than once. We should have seen it again; we looked for it 50 times.'* He forgot that our Arecibo message to the star cluster M13 in 1974 was also sent only once.

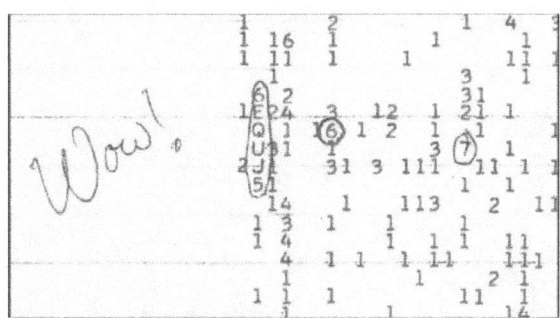

There was a comet in the same part of the sky that was emitting a radio frequency close to that being used. If this was the culprit, a natural source, then why wasn't it a constant signal that other radio telescopes could have observed?

Had the signal been confirmed as coming from an Alien Intelligence, it would have become the most momentous event in all of scientific history. It almost certainly influenced many movies from that point on; Close Encounters of the Third Kind, The Arrival, Species and Contact are just a few.

The 'Wow' signal is so well known that it deserved a display at the Science Museum near Atlanta, Georgia; the home of the Apollo 6 spacecraft. Photo by the author.

Photo by the Author

I visited the 'Big Ear' radio telescope in July 1992 (poser on the left) and made friends with the volunteers that operated it at the time. During the morning this picture was taken, a NASA representative approved a loan of a complex receiver worth $1million specifically designed to receive artificial signals from deep space. They asked me to blab to the rep as if I was from Cambridge Radio Observatory. The experience was fun but also learned how science is carried out at this level. This unique site was dismantled in 1998 for a golf course.

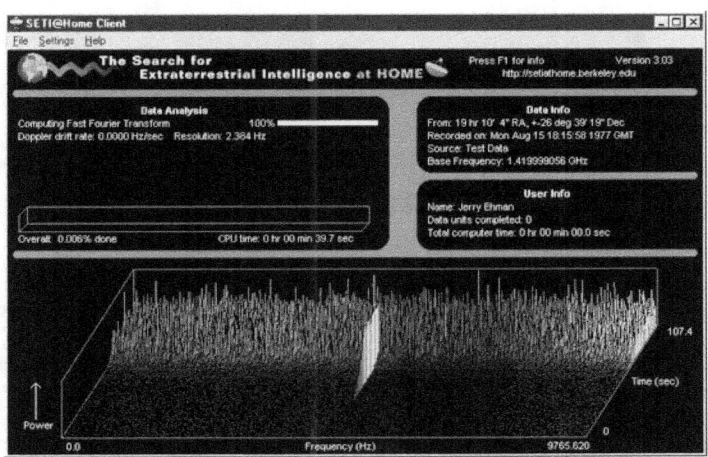

This is how the 'Wow' signal would have looked on the SETI data screensaver – see chapter on 'Discoveries on your computer.'

153

Chapter 32   Frequencies

If a civilization did develop technology the same way we did, it would probably change and advance rapidly. Radiating large amounts of power into space is very wasteful, so over time civilizations like ours would probably try to find methods that do not involve broadcasting signals everywhere. TV signals as an example of this. Broadcast TV was going strong from the 60s until recently, but now everything going over to cable, which cannot be detected from a distance. So to detect our TV transmissions, aliens would have to be specifically looking our way in the right 40-year time period, or they would pretty much miss our 'strong TV signal' period of evolution.

There are millions of stars for us to look at, and so searching for another civilization's random broadcasts does not seem too efficient. This random radiation really is not "free" to a civilization either; it is a costly service. In addition, higher frequencies such as Light do not travel well for long distances. It is easily scattered by small particles like interstellar dust. The SETI researchers look at long wavelength radio signals for a couple reasons. First, radio waves travel straight through the large amounts of dust in our galaxy, and second, hydrogen (the most common element) radiates at around 1420MHz. Any technologically advanced civilization that wants to study the galaxy will be looking at this wavelength and understand its importance, so if you want to contact another civilization, sending signals at or around this wavelength could attract attention. Cell phone signals from across the galaxy would never make it to us, and once again, they might only be using this technology for a very short time.

Another problem is that signals have to be strong for us to detect them. Most emissions from Earth (the exception is some radar transmissions) would not be detectable with our current systems at the distance of the nearest star. There is a lot of "noise" from signals on Earth which makes it difficult to detect faint signals. SETI routinely finds "signals" that are produced on Earth or in Earth orbit. In fact, there are only a few frequency bands dedicated to any type of astronomy, and outside these frequency ranges (sometimes even inside them) there is so much interference from our communications that detecting faint signals is impossible.

Finally, to address the expense issue, I would first like to point out that taxpayers are not currently paying for SETI! This funding was eliminated many years ago and private donors now fund the enterprise. We on Earth have not tried very hard to send signals out into space. It is entirely possible that civilizations will decide not to waste power and energy in an effort to contact others. However, a dedicated broadcast will be the easiest transmission to pick up!

As technology develops, we will probably have a better chance of detecting random signals. Someday if we are able to closely view extrasolar planets, we might be able to tell if they have emissions at other parts of the spectrum, but for now people are trying to do the smartest thing that technology and money will allow. There is no reason to believe that there is not a civilization that would want to try to contact others across the galaxy via radio. Universally, radio is a low cost, highly penetrating range of frequencies.

Part of the Cambridge Radio Telescope Array, UK.

Chapter 33   The Martian Fossils

In 1984, a scientist found a new type of meteorite in Antarctica and was given the name ALH84001. Months of research concluded it came from Mars. They took a small sample of it, heated the material and examined the trace gases that were released by 'bubbles' of air that was trapped inside. An asteroid must have struck it millions of years ago and blasted pieces of Mars off its surface and into flew into deep space orbiting the sun. Many years later, it entered the Earth's atmosphere and landed near the South Pole.

In 1996, NASA and the White House made the explosive announcement that the rock contained traces of Martian bugs. Photographs were released showing elongated segmented structures that appeared lifelike.

The excitement did not last long. Other scientists naturally questioned whether the meteorite samples were contaminated while on Earth. They also argued that heat generated when the rock was blasted into space might have created mineral structures that could be mistaken for microfossils instead.

A new investigation was conducted in 2014 using high resolution electron microscope techniques which were not available then. News of the findings was leaked to the news website 'Spaceflight Now.' Dr Kathie Thomas-Keprta and members of the original team at the Johnson Space Centre in Houston led the research. They claimed to have found evidence of fossilised life in the meteorite after all.

It focused on a more detailed analysis of tiny magnetic particles and carbonate discs within the rock. Certain bacteria on Earth are known to contain magnetite crystals, which they are believed to use as tiny compasses to help them navigate.

These crystals form unusual shapes when associated with bacteria which can be seen in ALH 84001, it is claimed.
In addition, the scientists say the chemical purity of the features they studied points to biology rather than geology, and a possible association with water.

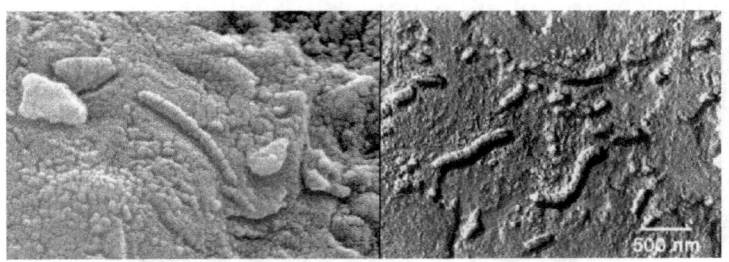

Dr Dennis Bazylinski, from the University of Nevada at Las Vegas, who reviewed the discovery, submitted an article in the Geochemical and Meteoritic Society journal;

'I think the paper is really excellent. I work on magnetic bacteria, and one indication there was life on ancient Mars is these particular magnetite crystals in the meteorite that look like they came out of magnetic bacteria. At first, I thought there might have been an error. I have no doubt about that now. I know there is no error. 'The big question is can these things be reliable magneto fossils, and that is a matter of debate. But it turns out that the magnetic bacteria make some very unique shapes of magnetite crystals. And one of the organisms we work with on Earth makes particles that look virtually identical to what we see from Mars in the meteorite.'

This does seem to be a powerful case for believing the meteorite carries genuine traces of past extra-terrestrial life. Some palaeontologists (fossil experts) claim that if you pick up any rock on earth, the chances of it having fossils inside is incredibly remote, so a similar rock being found from Mars containing fossils is extremely unlikely.

However, that point is assuming that the process of fossilisation on Mars was as uncommon as on Earth. Mars does not have the same erosion processes as Earth for a start. There are no plate tectonics on Mars and there would be weak tides in any seas and oceans. It is possible that creatures on Mars had a far higher chance of being fossilised and surviving for hundreds of millions of years than on earth. So therefore the statistics of picking out a rock at random on Earth or Mars will have different chances of discovering a fossil inside.

Over forty Martian meteorites have now been recovered as of 2016. Scientists are currently examining some of these for similar potential fossils. Better still, NASA rovers on the surface of the planet have been searching for fossils too.

The rover – Spirit at Columbia Hills on Mars, took this image. It may look like fossilised lichen. However, it turned out to be marks made by the brushes in the centre of a drill machine on the rover itself.

Chapter 34   Interstellar Travel

'Space is big. You just will not believe how vastly, hugely, mind- bogglingly big it is. I mean, you may think it's a long way down the road to the chemist, but that's just peanuts to space.'

Douglas Adams, The Hitchhiker's Guide to the Galaxy
English humourist & science fiction novelist (1952 - 2001).

If we or other beings are to travel amongst the stars and potentially meet other intelligent beings, then special forms of propulsion are required. If we find this impossible due to the restricting laws of physics, then we already have the answer that aliens cannot visit us. Nevertheless, this is not what is found.

To give some idea of how far the stars are, we need to compare the distances within our own solar system with the stars themselves. It took Voyager 2 twelve years to reach Neptune, nine years for New Horizons to reach Pluto in July 2015. At a speed of 52,000mph, the velocity of New Horizons, it would still take 54,400 years to reach the nearest star – Proxima Centauri; just 4.2 light years away... *Houston we have a problem.*

The record for fastest launch velocity belongs to the New Horizons probe that lifted off in 2006 to Pluto. This 478 kilogram, piano-size spacecraft sped away from the Earth at a blistering 36,000 mph (58,000 km/h). As it flew past Jupiter, the gravity from it accelerated up to 52,000mph.

The Juno spacecraft now at Jupiter is currently the fastest human-made object, it remains to be seen how long the space probe can hold onto the impressive title. Solar Probe Plus, a NASA mission scheduled to launch in 2018, is designed to fly into the sun's atmosphere. Due to the enormous gravity of the sun, the probe is expected to reach velocities as high as 450,000 mph (724,000 km/h). At this speed, you could travel from the Earth to the Moon in 30 minutes.

Here is a list of ideas that that could get probes and even humans to the stars. These in theory work within the known physical laws followed by a few fanciful ideas that just may be brought into reality providing the existing laws can be 'expanded.'

Project Daedalus: A detailed report was published in 1981 of a star-ship powered by pellets of Helium 3 that undergo Nuclear Fusion via a laser imploder... phew.

Experts through the British Interplanetary Society compiled the report. It included mathematical equations, chemistry, materials required and its potential velocity. It could theoretically allow a two-stage probe to reach the nearest stars within 50 years. This became the first serious study into this possibility based on near-future technologies.

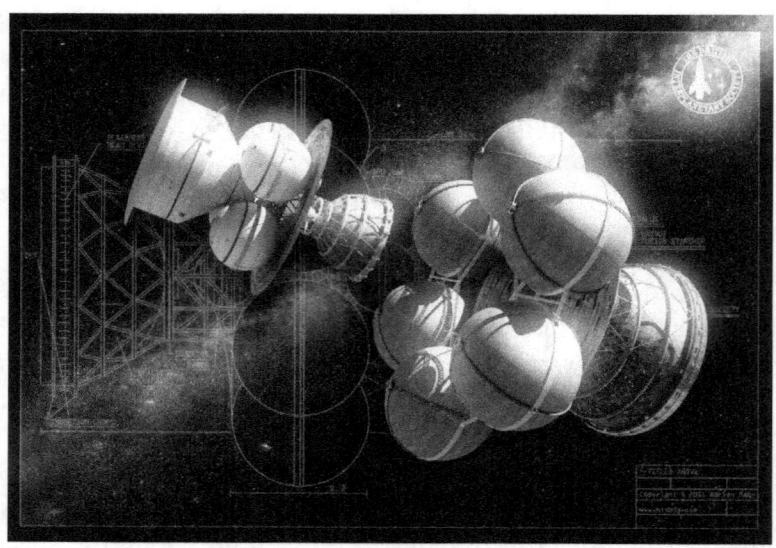

Generations in Space: Spacecraft can currently be accelerated up to around 55,000mph at a push. At this velocity it would still take around 1,300 generations; adults will have children who become parents and so on 1,300 times or so. Long before they

160

have reached a target star system, a more advanced craft will overtake them within a few decades let alone 54,000 years. This will be realised before the project even leaves the drawing board, so would the effort to build such a ship be made?

The numbers and enthusiasm changes if a much more rapid velocity is achieved. Even a 0.5% of light speed (3.3 million mph) will make a massive difference. The nearest stars could be reached in 900 years or around 22 generations. The occupants would need some form of gravity. Without delving into the weird physics shown on Star Trek that may never be possible, there is a way.

Fill a bucket with around 2 inches of water. Go outside and begin to swing it back and forth. With each swing, make the arc longer and longer until you feel ready to completely swing it over your head. If timed correctly, you will not be drenched. The water will remain on the bottom of the bucket even when it's upside-down. This is due to Centripetal Force - a false form of Gravity. If you could be in that bucket, then you too could stand on the bottom the whole time and not fall out.

If a large ring can be built in space and was spun, then the same force is created. At a particular speed, a force equal to the Earth's gravity can be sustained for years. Momentum will keep the force constant.

Movies such as '2001 A Space Odyssey', '2010 Odyssey Two' and 'Interstellar' all correctly show this principal. The very first time this was demonstrated in space for real was a crew on board the Skylab Space Station. A ring of cupboards were built around the inside of it that produced a neat circle. The astronauts ran around it and generated a little 'weight' of their own. The Mir Space Station was spun a little once and the crew rested against a far end module and felt a small 'G force.'

Below; a recently proposed inflatable ring that can be spun up to produce artificial gravity on the International Space Station. In theory, this could produce a full reproduction of the Earth's gravity. This one single move could prove the concept of not just living in space for up to a year at a time, but an entire lifetime if required in full good health.

Space travellers can live their entire lives in space and remain healthy by using this technique. If several thousand people can live in a massive habitat with its own energy supply, water, air etc. then there is no reason why such a vehicle cannot accelerate to a speed of over 3,000,000 mph and head for the stars. All will have a good quality of life, but their descendants will reach a new planetary system and make a new home there. Arthur C Clarke's book, Rendezvous with Rama suggested a combination of a rotating space station and Suspended Animation for travelling between stars. The movie 'Elysium' 2013 with Jodie Foster and Matt Damon showed a similar rotating space station.

Suspended Animation: In the movies, suspended animation involves freezing people solid (or nearly so), then thawing at some point in the future when new medical advances have taken place to treat their conditions that may be incurable today or for a long duration spaceflight. An emergency preservation and resuscitation (EPR) technique is not quite so extreme, but it will reduce body temperature to 10° Celsius by inserting cold saline into the blood system. This will slow the blood flow, which will prevent the body from bleeding out of any wounds. The low temperatures will also slow other aging processes.

The human body can currently sustain this state of hypothermia for about two hours. While this is not a long-term state as shown in movies, this could easily provide enough time for surgeons to perform emergency lifesaving surgery. Trauma patients who suffer cardiac arrest have an extra 7% chance of survival, giving this technique has some very real and amazing implications.

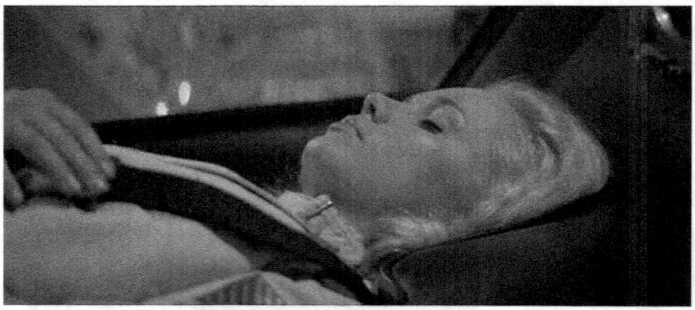

A scene from 'Planet of the Apes.'
(She had a very small part to play - unfortunately).

Dr Peter Rhee first tried this method on 40 pigs in 2000, with the results published in 2006. Perhaps a more advanced form of this technique can be used in Spaceflight. Instead of using valuable oxygen, water and food and becoming extremely bored, the astronauts can be frozen and automatically resuscitated years later when the craft nears its programmed destination.

Automated Galaxy Exploring Probes: Instead of the concept of living creatures exploring the Galaxy, we ourselves currently send out probes to planets, asteroids, comets, moons etc. These are cheaper to manufacture than manned spacecraft, don't need life support systems, or get bored, nor do they need to return home.

Hungarian born John von Neumann, (1903-1957) who worked on the Atomic Bomb in WW2 and early computer science, proposed the concept of self-replicating probes travelling from one planet to another, reproducing itself from existing materials at the destination. They then send out the new probes with updated programs to other worlds and so on. Discoveries made on the journey by each probe may be transmitted back to the home planet.

If a velocity of around 40% light speed was achievable for such vehicles, then it would be possible for one species to explore every 'corner' of the Galaxy within 4 million years without leaving their home planet. However, the evidence of their presence would be galaxy-wide with all those millions of probes on countless millions of planets and moons transmitting trillions of Giga Bytes of data across the Cosmos. So once again, where are all these probes and signals? Perhaps it's these we may detect first any time soon.

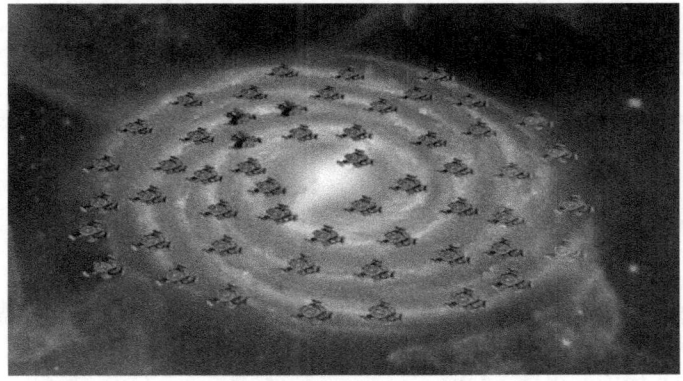

Self-replicating probes can be sent out to explore the Galaxy instead of leaving home;

165

Almost Light Speed: Albert Einstein (1879-1955) wrote a paper entitled 'The Special Theory of Relativity. It concerned with what happens at extremely fast velocities. Newton was correct with his laws of motion except that at unimaginable high speeds, unusual effects come into play; these we not predictable or testable in the 1600's.

The one single point we should understand is that the faster one travels, the slower time flows for that traveller. This is a very real effect; not science fiction, and has been tested billions of times for over a century. The old TV sets for instance used a 'Cathode Ray Tube.' Electrons were fired out of an emitter at a fraction of the speed of light. Time for those electrons slowed and was affected more by magnetic fields that created the picture than they would under Newton's laws of motion. Therefore, every time such a TV set was turned on and the picture correctly formed on the screen, it was confirming Einstein's Relativity.

If a spacecraft could travel at perhaps 90% of the speed of light (186,282.3959 miles per second), then the entire craft and any occupants would experience time slowing to 43.6% of normal speed on Earth. Small 'c' is the speed of light as a mathematical expression.

% of c	Dilated to %		% of c	Dilated to %
10%	99.5%		60%	80.0%
20%	98.0%		70%	71.4%
30%	95.4%		80%	60.0%
40%	91.7%		90%	43.6%
50%	86.6%		99%	14.1%

To save time with the calculations, this table demonstrates the relationship between the slowing of time and velocity.
The following formula is the heart of Time Dilation as it became known. T_A is the time observed by an observer with respect to the object, T_B is the time measured by the object.

166

$$T_A = \frac{T_B}{\sqrt{1 - \dfrac{v^2}{c^2}}}$$

It may take many years to travel to a star, but if the craft travelled fast enough, then the time experienced on board would be much less. By using the previous table, it becomes calculable that at 99% light speed the crew can travel to a star say 20 light years away but only experience 2 years and 9 months; not including acceleration and deceleration. At least this becomes more manageable and bearable. At beyond, 99%, the effect takes hold in a big way. A trip to a very distant star of thousands of light years can be reached in 10 years according to on board clocks and heart beats.

This point allows humans to travel to the stars well within a single lifetime. The same effect will allow other species to reach us too or at least colonise vast parts of the Galaxy. We have no idea how to produce so much power for one vessel to accelerate to almost 186,000 miles per second. This is the situation that Sir Isaac Newton faced when he could not comprehend how we could send people to the moon; at least he knew it could be done using his laws of motion.

The greatest disadvantage though is regarding another section of relativity; the higher your velocity, the greater your mass becomes too. A moving high-speed train will weigh a gram or two more than the same train at rest by a station platform. Again this is a real effect; not fiction. The effect becomes much more acute at higher velocities compared to light. A moving 1000-ton spacecraft at 99% light speed will weigh 1071 tons. More energy would be needed to accelerate further.

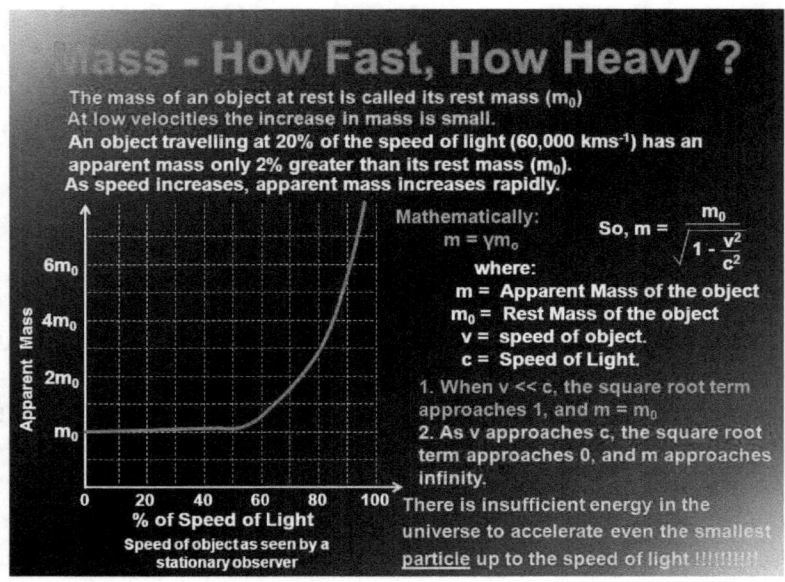

Mass - How Fast, How Heavy ?

The mass of an object at rest is called its rest mass (m_0)
At low velocities the increase in mass is small.
An object travelling at 20% of the speed of light (60,000 kms[-1]) has an apparent mass only 2% greater than its rest mass (m_0).
As speed increases, apparent mass increases rapidly.

Mathematically:

$$m = \gamma m_0$$

So, $m = \dfrac{m_0}{\sqrt{1 - \dfrac{v^2}{c^2}}}$

where:

m = Apparent Mass of the object
m_0 = Rest Mass of the object
v = speed of object.
c = Speed of Light.

1. When v << c, the square root term approaches 1, and m = m_0
2. As v approaches c, the square root term approaches 0, and m approaches infinity.
There is insufficient energy in the universe to accelerate even the smallest particle up to the speed of light !!!!!!!!

Apparent Mass (y-axis): $6m_0$, $4m_0$, $2m_0$, m_0

% of Speed of Light (x-axis): 0, 20, 40, 60, 80, 100

Speed of object as seen by a stationary observer

If the table following is studied carefully, you can see that the power required to travel at very close to light speed becomes ridiculously high. Achieving light speed itself would consume an infinite amount of energy; more than the universe can ever produce.

c %	Mass increase %	c %	Mass increase %
25	1.033	99.9	22.37
50	1.158	99.99	70.71
75	1.512	99.999	223.6
95	3.203	99.9999	707.1
98	5.025	99.99999	2236
99	7.089	99.999999	7071
99.5	10.01	99.9999999	22361

Therefore, nothing made of atoms can ever travel at light speed or especially beyond. There may just be away of sidestepping the rule...

Wormholes: So far, we have investigated methods of exploring the galaxy with known laws of physics. What about the unknown? Okay we can speculate and make anything up we like; it's unlikely ever to materialise. Alternatively, we can just push the known laws of the universe just a little and see what we find.

We already know that black holes exist. Without dedicating several hundred pages of this book on how they form or behave, scientists are sure that if other dimensions exist, this could be a way in. As the nearest known black hole to us is over 1000 light years away, it would take hundreds of generations just to get there let alone find out what is possible and what is not as you dive in.

By studying black holes and combining the knowledge gained from atom smashing machines (particle accelerators), evidence is slowly growing that other dimensions do indeed exist. It may be possible to create a 'hole' in space, float into it and reappear hundreds of light years away in a controlled fashion and on demand whenever we want.

If we are being visited today by alien beings, then this could perhaps be the only realistic form of rapid travel between stars. It is inconceivable that they take thousands of years to get here or use up enormous amount of power just to create a crop circle (joke), study us a little and move on to another star system or head back home. Easy regular access would be essential.

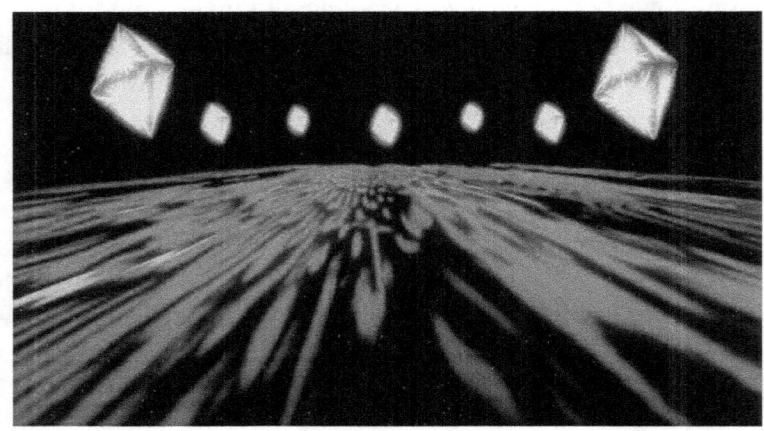

The movie '2001 A Space Odyssey' is the first to demonstrate the wormhole theory. Other such movies include 'Stargate', 'Contact' and 'Interstellar.' These propose that technology can create a wormhole or travel to an existing one built by aliens as an instant gateway to other planets. So far, there is currently nothing in physics that would disallow such a form of travel.

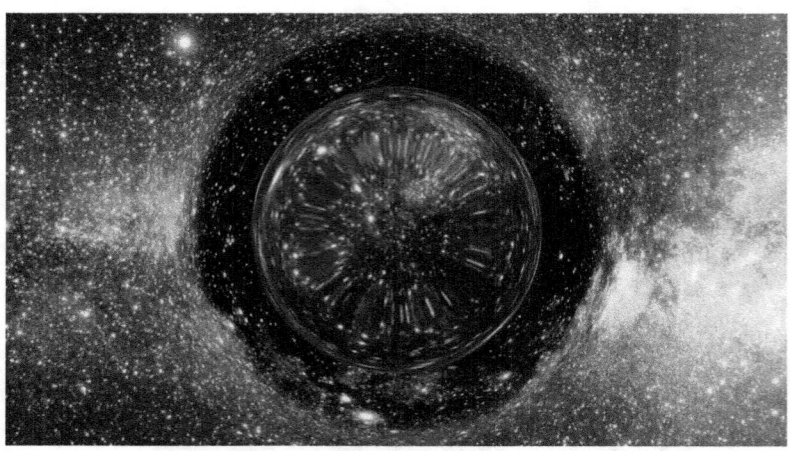

Natural multi-dimensional gateways into a form of hyperspace may well be the ultimate form of travel. These would distort space around themselves just as black holes do; but without that messing around with super-strong gravity.

The purpose of this chapter is to show what effort is required to travel between stars. It is certainly not impossible, but tricky to say the least. It proves beyond doubt that any civilisation can adopt any one or more of these techniques to travel from one star system to the next and occupy the entire galaxy. Computer simulations show that within an estimated time of 1 to 4 million years is all that is required. They may not set out that as a target but would be a natural progression instead; just as we have spread across the globe over several hundred thousand years from our humble beginnings in Ethiopia, Africa.

This scenario almost certainly has not occurred in our galaxy, their signals by now should have been picked up somewhere in the sky. Better still, why didn't one of the colonies discover Earth 10 million or 100 million years ago and settle here? All our resources are in place and no remains ever found anywhere in the solar system so far. In addition, why leave? The Sun and Earth are both very stable.

This is the heart of the 'Fermi Paradox. If extra-terrestrials exist at all, why haven't we detected them already? Perhaps we are doing something wrong.

Perhaps technological based civilisations are destined not to last very long. This could be the sad answer to the Fermi Paradox. All it would take though is one to break out of this trend and colonise the Milky Way in an unstoppable flow.

Chapter 35 Dyson Spheres

A Type II civilisation should be able to construct rings of solar panels around their parent star. If so, as it rotates across our line-of-sight, the star should dim down dramatically then brighten back up on a regular basis. A planet passing in front of the star will dim the star too but in a much different way. The Kepler Telescope was perfect to discover such phenomena. As of 2016, two stars performing such dips in starlight have been discovered.

KIC 8462852 is known as Tabby's Star, after Yale astronomer Tabetha Bovajian. In September 2015, she published first details of its strange behaviour on a Planet Hunters Website. This service invites people to scour astronomical data in the search of new exoplanets; Boyajian noted that the otherwise normal output of the type F3 star displayed an odd series of ten substantial dips in brightness.

These dips were unlike anything seen elsewhere in the universe. A Jupiter-sized world passing in front of an F-type star would cause a dip of about one per cent, lasting a few hours and repeating every orbit, often just a few days. The first recorded dip in brightness of Tabby's Star was about one per cent, but lasted a week.

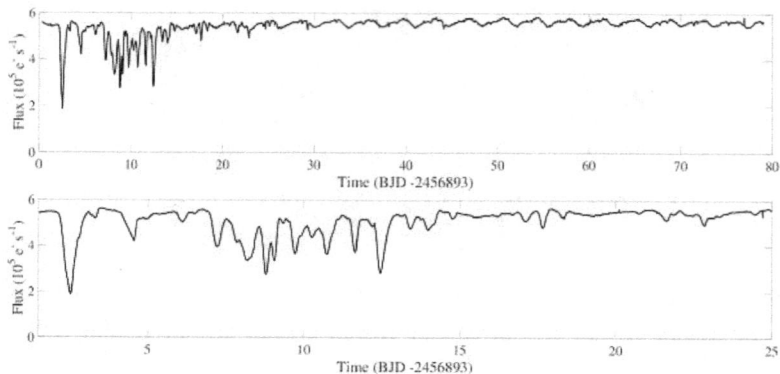

It is conceivable that rings of energy absorbing material orbiting the star cause the observed dips in brightness. This energy can be beamed back to a planet anywhere in the star system and converted into electricity.

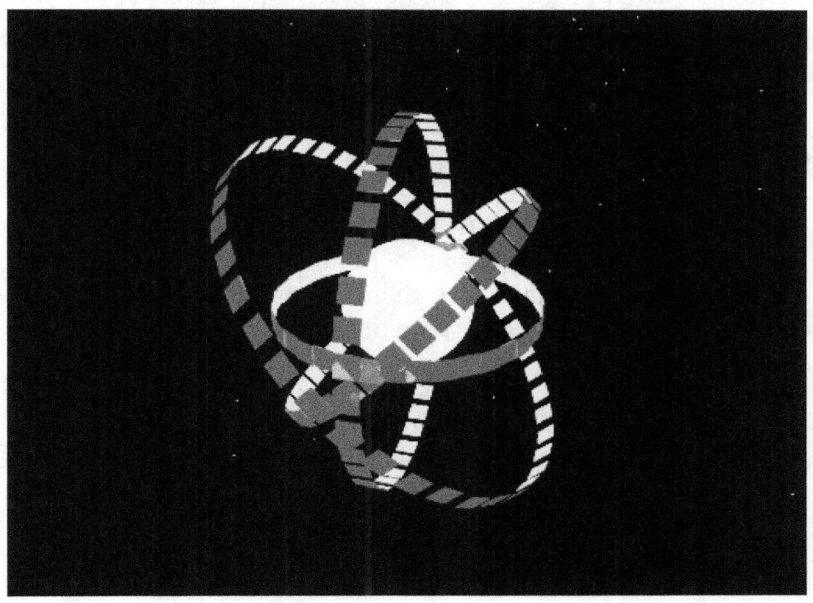

A Dyson Sphere first proposed by physicist and astronomer Freeman J. Dyson first explored this idea as a thought experiment in 1960. One ring is built first, then others can be added later to eventually encapsulate the entire star as absorb its entire energy output.

The observation that really got people taking notice was the "swarm" of ten dips in 2015 that defied explanation. They were close together, but irregular in shape, variable in depth from a fraction of a percent to over 20 per cent, and with no obvious timing pattern.

The deeper dips would require something like 1,000 Earth-sized planets to pass in front of the star at the same time. When Pennsylvania State University astronomer Jason Wright suggested in October 2015 that an alien megastructure was one possible explanation for this erratic behaviour, the internet was flooded with theories.

While such disks would emit infrared light, an alignment would make this radiation invisible to us due to the closeness of the star to it. So far, not all the evidence discounts a Dyson Sphere.

Young stars are known to have similar large dips in brightness that this star has, and they have recently been shown to come in a variety of inclination angles too.

A second possible star showing the same signs is called EPIC 204278916. This star experiences the similar dips to those seen on Tabby's Star, and is less than 11 million years old (not enough time to allow for life to evolve) with a mass much lower than our Sun.

This led to the SETI Institute turning its instruments to these stars, but found no artificial radio signals yet. Other astronomers rushed to offer alternative explanations, including a cluster of comet fragments or a debris field created when large asteroids or planets collide.

A cluster of dark comets could produce a silhouette giving a similar data readout as a Dyson Sphere; NASA.

Chapter 36   Planet K2-18B

Possibly the most exciting exoplanet discovery to date, has the romantic name of K2-18B. It lies in the constellation of Leo the Lion and is 124 light years away. The planet orbits a red-dwarf star every 33 days and perfectly situated with its habitable zone. It was discovered in 2015 by the Kepler Space Telescope.

The next event was in 2019 with the discovery of water vapour within its atmosphere. One of the first ever detected. Excitement amongst the world's scientists grew. Further observations in 2023 uncovered carbon dioxide and Methane. It was looking more like a water-based planet with a possible hydrogen-rich atmosphere.

Measured diameter comparison. Credit ESA.

In 2025, the James-Webb Telescope made the most fundamental discovery of all, Dimethyl Sulphide (DMS), a chemical that could serve as a strong signal for life. Micro-organisms on Earh produce it. The quantity of DMS is around 20 times that of Earth. This material is always short lived, and so it must be being rapidly replenished. A life bearing planet would account for all these points. However, some scientists can point to lab experiments that can produce DMS chemically / no life. Shucks!

If the next step is 'Let's go there and explore it', the fastest spacecraft leaving the Solar System is Voyager 1. Its current velocity is 16.9km/second and at that rate, even if it were heading straight for K-2 18B, it would take 2 million years to reach it.

As the planet orbits in just 33 days, it often passes in front of the parent red-dwarf star. Observations therefor are regular and data builds rapidly. As new techniques are employed, evidence for life should build also. Again, we have to remain cautiously optimistic but be realistic. So far this remains the most promising candidate for our ultimate quest. Just one confirmed planet with life is all we need to alter our entire perception.

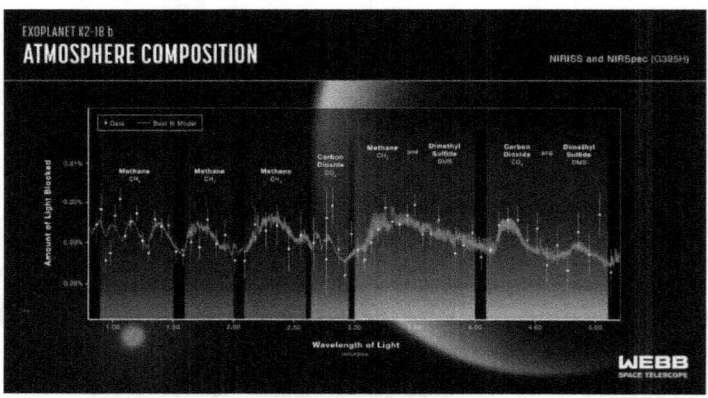

K2-18B is unlikely to ever become a second home for humans, the study's author, Dr Angelos Tsiaras, said... *"K2-18b is now the best candidate for habitability. It's the only planet outside our solar system that we know has the correct temperatures, atmosphere and water,"* the University of London (UCL)... *"At the same time, K2-18b is not a second Earth. This is a planet that is much bigger. It has a different atmospheric composition and it's orbiting a completely different type of star. The search for habitable planets is very exciting, but it's here to always remind us that this is our only home."*

The lead astronomer Bjorn Benneke of the University of Montreal, Canada, said... *"This represents the biggest step yet taken toward our ultimate goal of finding life on other planets, of proving that we are not alone."*

Section 3
What can we do?

Chapter 37   UFO Investigations

Anyone can become a UFO investigator. No qualifications are required but cannot promise you will be paid. Most are volunteers but a tiny handful has managed to perform this as a way of making a living. Nick Pope worked for the Ministry of Defence in the UK and received a regular wage. J Allen Hynek was the US equivalent working for the Air Force on Project Blue Book. Such jobs are rare. Anyone becoming a true expert on cases may become an author / lecturer and earn a wage on a self-employed basis; another available option.

To avoid embarrassment and being hammered by sceptics, a realistic and well-researched approach is essential. Do not always come down on the side of the believer. Learn subjects such as image processing, natural astronomical and meteorological phenomena and space flight. Knowledge of past cases and how they were investigated is essential.

A test of your own abilities would be to decide upon what a genuine conspiracy is and a false one. A true investigator will be able to discern one from the other by examining just a few points and asking the right questions. The Moon Landing Hoax is a classic example… they were genuine.

You may consider joining a UFO organisation such as British UFO Research Association (BUFORA) that even offers training courses. Another is Mutual UFO Network (MUFON) that also offer courses. The websites are found as links from the supporting www.outerspacebook.com site. The only point to make here is try not to be convinced that Flying Saucers are real by just reading about cases. Stories can be exaggerated over time, negative points may be omitted (such as a witness may have been drunk at the time or on something else), experts may claim to be experts but are complete amateurs instead. An investigator may come down on the side of the believer every time due to a personal endeavour to 'prove' these things are real. Every single aspect has to be verified.

All it takes though is one genuine case of a meeting with aliens along with indisputable evidence, and then we have solved the greatest riddle ever facing the human race. A cause worthy of some effort, but to this day, this has never been fulfilled. Perhaps you may be the first.

J. Allen Hynek (a paid UFO investigator) in the 1977 movie Close Encounters of the Third Kind; Columbia Pictures. I saw this eleven times at the cinema before it ever appeared on TV or video. Some scenes were based on actual reported events.

Nick Pope worked for the Ministry of Defence, UK. Although very knowledgeable, he tends to fall on the side of the believer far too much and too rapidly in my humble opinion.

Arizona Project: My wife, Amanda and I have purchased a 36acre plot of land in Northern Arizona. With a very low horizon and 300 clear nights every year, we are setting up an automated sky watching camera system. Any light source that

179

moves rapidly in any part of the sky will be recorded and emailed to us back in the UK. This should include bright meteors, satellites, ball lightning and other related phenomena. Most of the images will be known events; it's the unknown we are hunting for. The UFO question has not been answered fully. This low cost system may inspire other people to duplicate it and so a network of such cameras could provide a powerful scientific tool for various forms of research. Two similar cameras a few thousand feet apart will produce 3D information that can confirm exact height and direction of anything within the atmosphere or just beyond by means of a simple 'parallax shift' calculation from the two photographs.

The first camera platform completed April 2017. The low horizon continues 360° and no light pollution from cities. For further details... www.northstaroasis.co.uk

This site is under the flight-path of some vehicles from Edwards Air Force Base, Area 51 and Vandenberg Air Force Base. This combination produces very interesting skies.

As from April 2017, parts for the observatory have been constructed. The first camera platform is now complete and many photographs of the night sky have been taken to prove the concept of using this site for Astronomical research.

14 April 2017; on the very first moonless run, an unusual event took place and was recorded on four photographs. It was so unusual that the observation was forwarded to the Mutual UFO

Network – MUFON. It was given a case number and I was interviewed extensively. Photographic experts within the organisation are currently investigating the case as this book went to print. Two of the most interesting parts of the images are below. They appear on more than one picture so these are not anomalies but real. Updates will be published on **www.outerspacebooks.com**

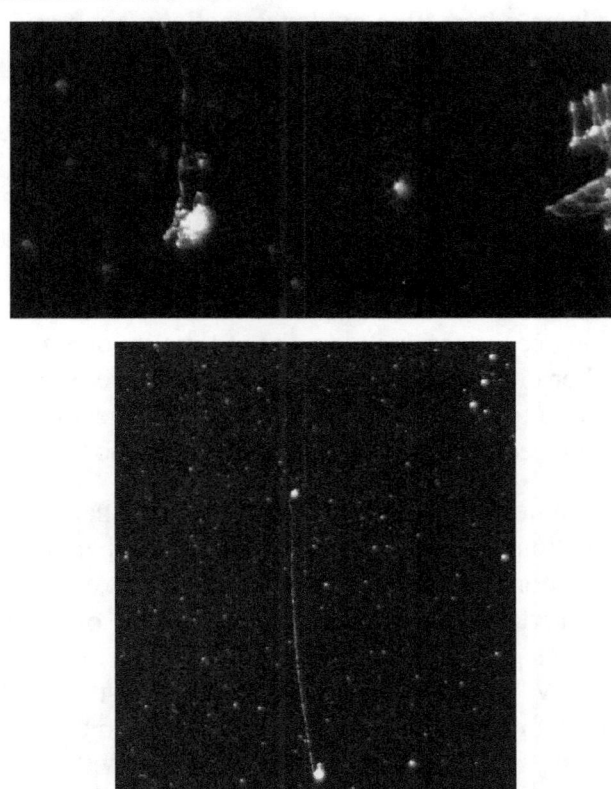

You may recognise Orion's Belt; three stars in a row. The bright star below is Sirius, but a 'tether' drifted past it, refracted the light, and produced, as far as I can analyse, a false image of Sirius at the end. Each of the four photos shows similar events across the sky for two minutes; very odd.

Chapter 38   SETI Projects

Search for Extra-terrestrial Intelligence (SETI)

If intelligent, communicating civilizations exist in the Milky Way, how can we learn that they exist? While there are millions of reports of UFOs, a thousand fresh sightings a month, to date there has been no credible evidence that any alien civilization has ever visited the Earth. Since the distances to the nearest stars are several light years or more, and our current technology only allows us to build ships that would require thousands of year to get to them, the stars are simply unreachable on a human timescale. Even assuming that other civilizations might be capable of building ships that can fly much faster, a round trip to a star 20 light years away is at the very least 40 years. Since physical travel between Earth and any nearby stars is improbable because of such lengths of time involved, if we are to find other civilizations in the Milky Way we expect it will be by communication using radio waves or other wavelengths rather than direct visits. This would be much more practical.

Light / Radio travels faster than any other means of communication, so a sufficiently advanced civilization may try to directly communicate using such low cost technology. Beyond direct, purposeful communication, our planet is broadcasting signals out into space every day in the form of our radio and TV broadcasts. When we broadcast radio signals around the world for you to listen to, those same signals also leak out into space, and so any civilization with a sensitive enough detector can receive say Laurel & Hardy shows from decades ago. By the same logic, if we try, we should be able to detect signals sent by a distant civilization for the same reason. However, the signal from a radio transmitter becomes weaker as it moves farther from the source, so the radio telescopes here or built by a distant civilization must have powerful specifications.

Radio waves take less energy to produce than say gamma rays; it is cheaper to generate radio signals than gamma-ray signals. So, we expect that radio waves are the most efficient way to communicate over large distances.

The Milky Way contains millions of objects that give off light from radio through to gamma rays, and so we want to choose a wavelength that will not be swamped by the Milky Way or absorbed as it travels through space.

Since we cannot know ahead of time anything about other civilizations that may be listening for signals from us or who are trying to communicate with us, the best that we can do is take educated guesses at how we might hear them.

Scientists who have been pursuing Search for Extra-terrestrial Intelligence or SETI research have been since the 1960s using radio telescopes to search for signals from other civilizations. These searches have concentrated on a region in the radio part of the spectrum. In a part of the radio spectrum where the emissions from the Milky Way and Earth's atmosphere is at a minimum, there is a wavelength associated with emission from Hydrogen (H) and another with emission from hydroxyl (OH). Since H+OH produces water, this part of the spectrum is referred to as the water hole. The assumption is that since this is a part of the spectrum that many astronomers already study and because the background is very low, it is a logical frequency for a distant civilization to try to communicate with us. Many of the SETI experiments that have been conducted have tuned their radio telescopes to this part of the spectrum.

The next question is, if astronomers have been searching the water hole for a signal from another civilization, has one ever been received? The answer is maybe! The "Big Ear" radio telescope detected the "Wow!" Signal as mentioned in a previous chapter. It has the appearance of a real SETI signal, but it was never able to be independently verified or repeated. In 1960, radio astronomer Frank D. Drake, then at the National Radio Astronomy Observatory (NRAO) in Green Bank, West

Virginia, carried out humanity's first attempt to detect radio transmissions from aliens. Project Ozma was named after the queen of L. Frank Baum's imaginary land of Oz -- a place "very far away, difficult to reach, and populated by strange and exotic beings."

The stars chosen for the first SETI search were Tau Ceti in the Constellation Cetus (the Whale) and Epsilon Eridani in the Constellation Eridanus (the River), some eleven light years (66 trillion miles) away. Both stars are about the same age and similar type as our sun.

From April to July 1960, six hours a day, Project Ozma's 85-foot radio telescope was tuned to the 21-centimeter emission (1420 MHz) coming from cold hydrogen gas in interstellar space. A single receiver scanned 400 kHz of bandwidth. The astronomers scanned the tapes for a repeated series of uniformly patterned pulses that would indicate an intelligent message or a series of prime numbers - 2, 3, 5, 7, 11, 13 etc.

With the exception of an early false alarm caused by a secret military experiment, the only sound that came from the loudspeaker was natural noise from stars and no meaningful signals on the recording paper either. After these first steps, systematic searches for other civilizations became a feasible and respected scientific project.

Several movies show this research method in various ways - Close Encounters of the Third Kind, The Arrival, Species and Contact.

The Paul Allen Array: The Allen Telescope Array (ATA), formerly known as the One Hectare Telescope (1hT), is a radio array dedicated to astronomical observations and a simultaneous SETI search. The array is situated at the Hat Creek Radio Observatory, 290 miles (470 km) northeast of San Francisco.

The project was originally developed as a joint effort between the SETI Institute and the Radio Astronomy Laboratory both at Berkeley, California, with funds obtained from an initial $11.5 million donation by Paul Allen co-founder of Microsoft). The first phase of construction was completed and became operational on 11 October 2007 with 42 antennas. After Paul pledged additional funding to support the construction of the second phase.

Although overall Paul has contributed more than $30 million, it has not succeeded in building the 350 x 6.1mt dishes originally conceived, and the project suffered an operational nightmare due to funding shortfalls in 2011.

Subsequently, UC Berkeley exited the project, completing divestment in April 2012. The facility is now managed by Stanford Research Institute, an independent, non-profit research body. As of 2016, the SETI Institute performs observations with the ATA between the hours of 6pm and 6am each day, 7 days a week. Financial problems constantly plague the project.

Chapter 39   Discoveries on your PC

Millions of people together, searching for a signal from the stars, SETI@home is by far the most successful public participation project ever undertaken in history! As of 2016 more than 12 million people in over 200 countries have taken part in the experiment, each contributing their computer processing time to the search for a signal from intelligent beings. Since its 1999 launch the project grew rapidly. However, had it not been for the timely efforts of The Planetary Society, SETI@home would never have got off the ground.

SETI@home is the answer to one of the most persistent problems in the search for extra-terrestrial intelligence: how to process the masses of data collected by radio SETI searches? Giant radio telescopes scan the skies, detecting radio emissions with increasing sensitivity and wider bandwidths and record all of them. Quite possibly, somewhere within all this radio noise there also a signal sent by intelligent beings' light years away for their own communications or specifically for us to hear.

Unfortunately, finding this signal in the enormous haystack of data requires almost unlimited computing time even on the largest and fastest computers on earth. SETI scientists, working with limited resources and even more limited computer time, usually have to severely restrict their analysis of the data just to arrive at any results at all. Sadly, this increases the risk that the true signal from ET, the signal that might be hiding in the data, will be completely overlooked.

In 1998, two researchers, a computer scientist David Anderson and SETI scientist Dan Werthimer, approached The Planetary Society with a novel idea. Instead of relying on strictly rationed time on one of the world's overbooked supercomputers, why not turn to a massively computing resource: the millions of personal computers sitting on desks and offices the world over, spending much of their time running worthless screen savers. If even a small portion of this computing power could be harnessed to process the SETI data then much of the computing problem would be solved at a stroke. It was a simple idea and they had the perfect name for it: SETI@home.

A sensitive receiver collects data at the Arecibo radio observatory in Puerto Rico, the largest radio dish in the world. The data is recorded on disks and shipped to U.C. Berkeley, California. A computer chops it up into thousands of standard-sized "work units," and makes them downloadable through the internet. Anyone with access to a personal computer is invited to download a simple SETI@home program and install it on his or her computer. The system evolved to employ a platform called 'BOINC'. This program automatically downloads work units from the web, and proceeds to process it whenever the PC is not being used. When the processing is complete, the program sends the data back to Berkeley and downloads new work units to process. Dual core computers can easily work on two units at once, a quad core, four at once and so on. Other computers can be added to a single account and the credit points add up.

Our personal score exceeds a million points between two projects. The process continues indefinitely, and the SETI data is slowly analysed. If ET is hiding somewhere amongst those signals, it will surely be found. If successful, the person that owns the home computer as well as the SETI project group share the discovery credit.

When Anderson and Werthimer approached The Planetary Society, the fate of the project was uncertain. For over a year they had been trying to raise money for their idea, approaching numerous companies in nearby Silicon Valley but coming up empty. The idea of involving the public in scientific research was a novel one and many found it difficult to believe that people would lend their personal computers to a SETI search for free. However, The Planetary Society reacted differently: SETI was in the Society's heart from the very beginning, and public involvement in science was at the core of its mission. After careful consideration, the Society donated $50,000 to help launch it. It also arranged for a similar grant from Paramount Pictures which was promoting the Star Trek" movie franchise.

SETI@home went online on 17 May 1999, and within a few months more than a million people across the world were processing work units on their personal computers, including myself that July. I had all three of my computers crunching data hourly and have done so ever since. I advertised the project in my planetarium lectures almost every time. Within a just year the number approached two million. No one had expected this explosive growth, least of all Anderson and Werthimer, who had dreamed of recruiting perhaps 100,000 people at most. With SETI@home, the search for aliens using real science had certainly captured the public's imagination.

The 300-metre-wide Arecibo Radio Telescope that had a receiver dedicated to SETI in the centre strapped to the overhanging cables. AS of November 2020, this particular dish has been decommissioned due to extreme storm damage.

A state of the art multi beam receiver at the heart of the giant 300mt wide / 1000ft dish had replaced the old and near-obsolete receiver at Arecibo. The program has been updated and searches for more types of signals with a greater sensitivity than ever before. On two occasions, the search's most promising signals have been gathered together, and a special search conducted to look for an alien signal coming from their direction. ET's call has not yet been detected as yet, but at this rate, if it is hiding somewhere in work units, it will surely be found any day.

China now has the largest radio telescope in the world measuring 500 meters in diameter; the radio telescope is nestled in a natural basin within a stunning landscape of lush green karst formations in southern Guizhou province. It took five years and £130 million ($180 million) to complete. Built in to the computer system are programs designed to find artificial signals from stars and assist the search for aliens.

The SETI project did something else as well that no one expected: it began a radically new way of doing science. The project's explosive success opened scientists' eyes to the fact that the public is eager to take part in innovative scientific research, and is happy to donate its computers and other resources for the cause. SETI@home was the first of a long line of scientific projects. Today the public is invited to take part in studies of climate change, cancer cells, protein folding, gravity

189

waves, interstellar dust particles, and many other science or mathematical problem solving projects.

Update for SETI@home; As of 2020, this facility has been taken offline. This was due to lack of donated funds, and the received data up to that point had been analysed. Once both points have been resolved, I'm sure the project will regain its full composure.

Download the BOINC software from our website link. Then you choose which project you wish to join. SETI is off course the one relevant for this book. As of December 2020, the SETI screensaver remains on hold while awaiting a new radio telescope and observing frequencies and funding. Check for updates, it could be running new research at any time.

I have also chosen the Rosetta project regarding the design of new molecules that can cure various diseases. The more you chose though; your computing time is diluted between each project. If you have a dual core computer, then one processor will work on at one project while the other works a second. If only one project is chosen, then both processors will work on two work units at once. A certain amount of credit points is given to you as a record of your total contribution. Sometimes a project may be suspended due to technical reasons, so your computer automatically works on the other project(s) instead. Your chosen registered name will always be attached to each unit completed to potentially give you the recognition for any discovery made.

Tasks		Transfers	Statistics	Disk	
Project	Progress	Status			Elapsed
SETI@home	88.197%	Running			02:11:01
SETI@home	45.319%	Running			01:48:43
SETI@home	61.638%	Running (0.0284 C...			02:28:34
SETI@home	23.998%	Running			01:45:37
SETI@home	23.748%	Waiting to run			01:43:42
SETI@home	0.000%	Ready to start			---
SETI@home	0.000%	Ready to start (0.0...			---
rosetta@home	0.124%	Running			00:01:16

Regarding the graphic display, choose the advanced display rather than the standard one. Many other options become available.

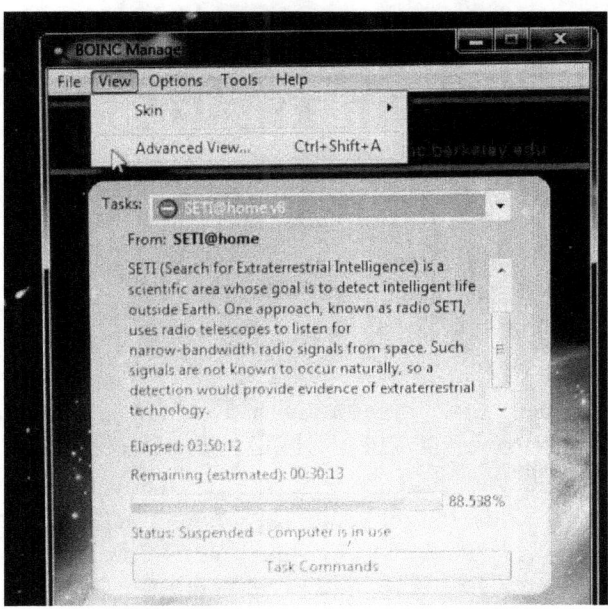

BOINC does claim not to slow your computer down if you use it when the number crunching is performing, I would challenge that. If you make heavy demands on your computer such as video editing, the computer will slow down. Under the

191

'Activity' section on the menu, you can select 'Run Based on Preferences.' Under the 'Options' menu the settings for 'Computing Preferences' can be found. This will allow you to control when the computing process starts and ends.

This can halt the BOINC processing all the while you are using the computer. As soon as you take a break, after 30 seconds or so, the processing will commence. Touch any key on the keyboard and it halts instantly. If you have a fairly fast machine and you are only working on something fairly simple such as an email, or a word document, then choose the 'Run Always' option. The computer will not run slow. It can be changed back anytime.

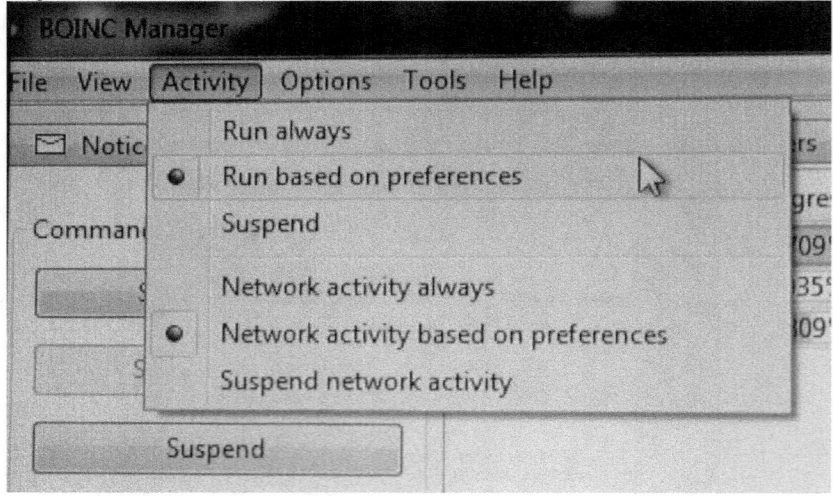

Many projects now operate worldwide on science and mathematical problems. The BOINC platform serves to control the data blocks, administrates your credit and the service is totally free. They never ask for many details for your registration and the info you give is never sold on. I have been a contributor since it first began in 1999 and have over 900,000 points for SETI as of 2019, and almost as much for Rosetta. Good luck with finding ET on your own computer.

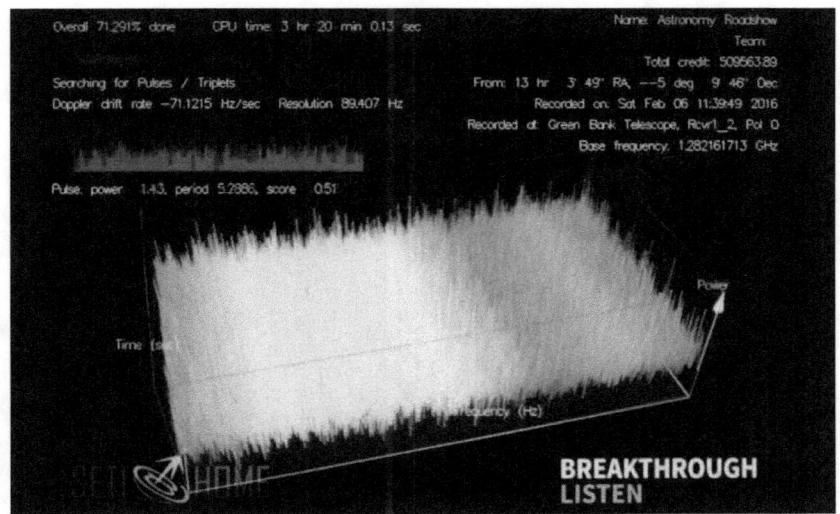

Planet Hunting: There is so much data coming from the Kepler telescope that astronomers do not have the time to sieve through all of it. The website www.zooniverse.org has compiled a simple tutorial and an application system for anyone to join the search. This was advertised on the BBC's Stargazing Live program in March 2016. *Direct link from* **www.outerspacebooks.com** Find 'Links' on the 'Are We Alone?' page.

Every 30 minutes, the Kepler Telescope monitors the brightness of 170,000 stars simultaneously for a sign of exoplanets via the transit method. When an exoplanet passes or transits in front of its parent star, the star shortly dims. This decrease in brightness is detectable by Kepler, lasts for a few hours or more, and repeats once per orbit per planet.

The time series of brightness measurements for a star is called a light curve. Automated computers are sifting through the Kepler light curves looking for the repeating signal of a planet transit, but there will be planets, which can only be found via the human ability for pattern recognition. At Planet Hunters, they are enlisting the public's help to inspect light curves and find these planets missed by automated detection methods. All you need is your keen eyes and a web browser to join the hunt.

It is just possible that you will be the first to know that a star somewhere has a planet, just as our Sun does.

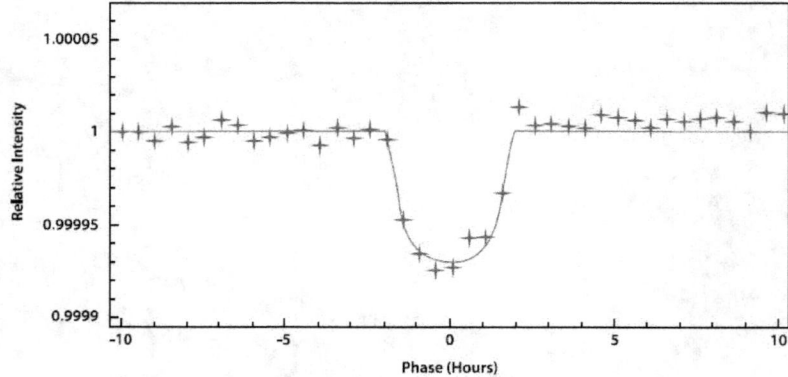

Chapter 40   Live Long and Prosper

A personal perspective

Dr Spock of the famed Star Trek TV series and movies often used the catch phrase 'Live Long and Prosper.' He was referring to an individual as a sign of good wishes to a friend. In this context, I am referring to the duration of the Human species as a whole. Dr Steven Hawking gave us 1000 yrs before our demise if we do not break away and set up self-sustaining colonies. There are many dangers that could eliminate us from existence while occupying one solitary planet.

Of all the 50 billion species from bacteria, microscopic multi-celled organisms to Humans over a complex history of 3 billion years of a living planet, intelligence with technology was perhaps an exception rather than an inevitable climax. Humans did not automatically produce advances in science. We were quite happy living in mud huts, hunting for food etc for tens of thousands of years before we even began to question the shape of the Earth or how big it was... and what are those dots in the night and why do they spin round each night?

Such thoughts commenced with the Greeks around 500 BC. Eratosthenes correctly calculated the earth's circumference, hence diameter around 250BC. The Roman Empire put a temporary halt to this and just gained power instead. Okay they did give us straight roads, toilets and baths, but not the big crucial thoughts. After the Empire's collapse, Europe sank into the 'Dark Ages' for centuries before people such as Copernicus, Galileo and Newton pulled us back out of this stale-mate scenario.

The era we are all living in today is also a very decisive one; more than most people are aware. The human population is growing at an ever-rapid pace. Our demands for materials and energy, is growing per person, decade after decade. We are also living longer, consuming more per lifetime as well as per year.

In some ways, we are rapidly developing a smaller world due to communications by phone and the Internet especially. Satellites and undersea optic fibre cables are at the heart of this revolution. In other ways we are becoming more divided and inward looking, this hides a fundamental danger.

With the major change in US administration, Trump has made a massive impression within NASA at various levels and may just give a hint of hope. I have maintained for many years that NASA has simply far too many rockets of various designs on their books. Each launch payload will have a list of possible methods of getting it into orbit. A much smaller listing of reliable rocket systems will enable the organisation to concentrate on perfecting what arsenal they have. In addition, he plans to almost abandon research within low Earth orbit; this can indeed now be carried out by other nations or private enterprise instead. Funds can then be transferred to possible manned deep space missions back to the Moon or even Mars. Potentially this move can relieve material and energy pressures on Earth through new technologies and discoveries.

Energy: Due to our massive demand for power, this is altering the very make up of our atmosphere. As we are aware, the climate is changing as a result. A massive global effort is required to solve this if we are ever to colonise space in a sustainable way.

China has built over 1300 new coal fired power stations, the filthiest form of electricity production. To keep people from panicking, they announce a new wind turbine or solar farm every now and then. We then tend to ignore the trillion tons of CO_2 they produce a year. The UK has seven coal power stations as of 2018 reducing to four by 2021. The USA was in the process of closing coal power stations, but with the new President, this will almost certainly be reversed. Donald Trump is well known for being taken in by Conspiracy Theories, one is that China 'invented' the Climate Change' scare. He chooses to ignore emission reduction agreements and expand coal production for new jobs; even though much of the coal industry is now mechanised and will create very few jobs in real terms.

He has a sharp business mind; this is probably the best chance the USA has in balancing the budget etc. However, with science, I feel he has a lot to learn. He claims that *'the climate has changed many times before and the world survives.'*

This statement is very true. Notice he never quoted a single number or proven example as a scientist would. Past climate changes often occurred over several million years; a formation of a new mountain range for instance. A very rapid climate change can occur due to a super-volcano eruption or a massive freshwater lake breaking out in an ocean. The recovery time in all such cases is in the order of several thousand to millions of years. Are we prepared to take a gamble that could take a geological timescale / hundreds of generations to solve?

Dust Storm in Phoenix, Arizona Aug 2016; by the author.

With all the major climate changes in the ice / geological record, the natural reason for most has been found. With the current issues, there is still no natural cause discovered. All our modern ground monitors, balloons, solar observations and satellites, nothing natural can be causing it. It has to be synthetic as it ties in closely with the increase in human-born

Carbon Dioxide emissions; Trump offers no explanation, no evidence, no charts, just a string of words. The climate does not care about politics, borders or finances.

Nearly half the species on the planet are failing to cope with the current global changes. According to an alarming new study, it suggests the sixth mass extinction of animal life in the Earth's history is now taking place.

A leading evolutionary biologist, Professor John Wiens, found that 47 per cent of nearly 1,000 species had suffered local extinctions linked to climate change with populations absent from areas where they had been found before. Wiens, editor of the Quarterly Review of Biology and a winner of the American Society of Naturalists' president's award, said the implications were serious because his review showed plants and animals were struggling to deal with the relatively small amount of global warming experienced so far. The world has warmed by about 1C above pre-industrial levels, it is expected to hit between 2.6 and 4.8C before 2100 if nothing or little is done to reduce greenhouse gases.

I have personally predicted since 1985 that false conspiracies along with material and energy shortages will affect the world in a chaotic way as it did in the 1930's; it has now happened. This is precisely why I produced a book and website to help individuals decide what real knowledge is and rather than the new phrase of *'alternative facts.'* Apply the same technique to Trump's claim of alternative facts on climate change and other issues; you obtain the real correct answer.

National debts of many countries are unsustainable. Money is printed from thin air, and people in powerful positions are burying their heads in the sand and ignoring such issues. They spend even more to keep everyone happy, make a country look rosy, and the votes come in. Climate change is already costing $trillions in increasing storm damage, coastal defences etc. and that does not include lost lives. Food production is becoming increasingly difficult to sustain worldwide due to harsher weather and various migrating pests are moving into new areas due to a changing temperature and moisture. This in turn forces research into 'designing' new kinds of crops via genetic engineering; producing its own health concerns.

Long Term: The tipping point to a complete disaster has not quite been reached. However, if one takes into account the lifetime of coal power stations, the potential lifetime of having someone negative in high political power, the time it would take to regain control of national debts - then a negative result of our experiment with the planet is looking much more likely.

The movie 'Interstellar' has an incredibly powerful message. There was a public firm belief in conspiracies such as the Moon landings hoax. Massive reductions in space exploration eventually became national policy. Other false beliefs around the world resulted in an inward looking human race; extreme climate change took hold, food production reduced across nations and a starving population arose globally.

The only solution was a top-secret space program; a wild mission to another galaxy, into a black hole and gain data on gravity that cannot be determined otherwise. Once the information was received on earth, massive space station structures were built very cheaply and with little energy launch millions of people deep into outer space. The climate on Earth would now have a chance to recover and a more sustainable future is possible for all humanity but in space.

This is just a story, sure, but if we continue down this path of this wild chance-taking experiment for much longer then I fear the worst. Perhaps this is an inevitable fate for all civilisations and could explain why we have not seen the Milky Way colonised by anyone. As mentioned at the beginning of the book, take away any personal feelings and just observe the facts to have any chance of answering our quest for others.

If we are indeed alone, there are no extra-terrestrials to chat to, then the billions of habitable worlds with living creatures in the galaxy (if they exist) are ours to explore and colonise. We should live long & prosper and take on this massive challenge if only we can think and act big enough.

At exactly the same time, our radio receivers and computing power has reached a point where if Extra-terrestrials are there, we should hear their signals literally any hour of any day and discovered by anyone including you!

Very Large Array, New Mexico; photo by the author,
September 1992

An artist's impression of a possible submarine mission to
Jupiter's Moon - Europa

Chapter 41 Other books by the Author

**New books will be published in the coming years. Refer to
www.outerspacebooks.com for signed copies, updated
listings and direct links. Most have a supporting website.
The following are amongst those available…**

There are B&W, Colour and e-book versions. All book sales
support four charities; Cancer Research UK, Kent Air
Ambulance, Smile Malawi Orphanage in Africa, British
Hedgehog Preservation Society

www.ingramcontent.com/pod-product-compliance
Lightning Source LLC
Chambersburg PA
CBHW070027210526
45170CB00012B/212